Generis

PUBLISHING

# A COMPUTER SIMULATION BASED ON A MARKOV CHAIN MODEL FOR THE DYNAMICS AND TRANSMISSION OF MALARIA

*NGOUABOU KOUAM Roger*

Title: **A COMPUTER SIMULATION BASED ON A MARKOV CHAIN MODEL FOR THE DYNAMICS AND TRANSMISSION OF MALARIA**

ISBN: 979-8-88676-210-5

Author: NGOUABOU KOUAM Roger

Cover image: https://pixabay.com/

Publisher: Generis Publishing
Online orders: www.generis-publishing.com
Contact email: info@generis-publishing.com

# Acknowledgements

I would like first of all give a special thank you to my father **KOUAM Maurice Calvin** for all his financial and moral assistance during all my studies. My thanks are also due to my supervisors

**Pr. L. P. FOTSO** and **Dr. William S. SHU** for exercising patience while guiding and directing me right to the end of this thesis. I greatly value their contribution to it accuracy and clarity.

My sincere thanks also go to all my lecturers, who gave me the background knowledge needed for the success of this work and did all to see me succeed.

# Résumé

Le présent mémoire a pour objectif de concevoir un environnement de simulation des dynamiques et de la transmission du paludisme. Tout d'abord, nous examinons comment une personne, à partir d'une piqûre de moustique peut être infectée. Ensuite, les différents comportements et interactions entre les populations hôtes et vecteurs de la maladie sont décrits et mis en évidence. Les populations concernées dans cette étude à savoir les hôtes (personnes) et les vecteurs (moustiques) sont subdivisées en compartiments. Chaque compartiment représente et décrit une des différentes phases de la maladie. Ces phases découlent du modèle SEIR de Ross-MacDonald. De cette étude préliminaire, un modèle basé sur le principe des chaînes de Markov est mis sur pied. La seconde activité de ce mémoire porte sur l'implémentation du modèle. Cette implémentation est faite par le biais d'un programme nommé «***Sipa***» (Simulation du paludisme). Les résultats obtenus à la suite de la simulation sont ensuite analysés, comparés et validés par rapport aux résultats provenant d'une version déterministe de notre modèle qui est stochastique.

**Mots clés** : Chaîne de Markov, Malaria, Modèle déterministe, Modèle stochastique, Modélisation, Population hôte, SEIR, Simulation, Vecteur.

# Abstract

This thesis deals with the simulation of the dynamics and transmission of malaria. First, we examine how a host (person) is infected through a malaria vector (mosquito) bite. Next, we describe the different behaviours and interactions between hosts and malaria vectors. The populations involved in our study (hosts and malaria vectors) are divided into compartments, which represent the different stages of the disease as given in the SEIR model of Ross-MacDonald. From this basic study, a Markov chain model is then developed. The second activity of this dissertation is to implement the model. This implementation is made through a program called «***Sipa***» (Simulation of malaria). The results obtained from simulation, which is a stochastic model, are analysed, compared and validated against a deterministic version.

**Key words**: Deterministic model, Host population, Malaria, Malaria vector, Markov chain, Modelling, SEIR, Simulation, Stochastic model.

# Table of contents

## List of figures

## List of tables

# CHAPTER 1: INTRODUCTION

Malaria is one of the most widespread vector-borne diseases worldwide. It affects generally people living in tropical zones and particularly in sub-saharan Africa. Unfortunately, almost all the countries located in these areas are either essentially poor or are developing countries and as a result, do not offer appropriate sanitation or health facilities. The lack of housing infrastructures and adequate provision for drainage often lead to the proliferation of breeding sites for the malaria disease vectors. Nowadays, the literature on dynamics and transmission of malaria is very large. Most of the researches are based on mathematical models (Aron and May [1], MacDonald [6], Ngwa[8], Ngwa and Shu [9] and Teboh[11]) or biological surveys; (Titanji et al. [12], Wondji et al. [13]). Whereas mathematicians model the dynamics and transmission of malaria as a set of mathematical equations, biologists work directly on human and mosquito populations. In spite of all these differences the main purpose remains the same: Eradication of malaria. For the same reason, the computer scientists build the virtual environments for the simulation of the dynamics and transmission of malaria.

In fact, the systematic control of malaria started after the discovery of malaria parasites by Leveran in 1889 (for which he received the Nobel price for medicine in 1907), and the demonstration by Ross in 1897 that mosquitoes are the vectors of malaria [7]. From that discovery, there have been many attempts to produce models that explain the dynamics and transmission of malaria. The mathematical model of Ross and MacDonald [6] has served as the basic hypothesis of many researchers. Using the Ross-MacDonald's works and the Markov chains, this project aims to build a virtual environment, called ***Sipa*** for simulating the dynamics and transmission of malaria.

A brief outline of this thesis is as follows: chapter two gives a background on simulation and modelling concepts, defines and describes malaria dynamics and then presents the basics of the SEIR model of malaria dynamics. Chapter three presents a full analysis of the model. In chapter four, the model is implemented and the results obtained are analysed and validated. Chapter 5 is the conclusion. Appendixes A, B and C contain respectively the algorithms, the initial values and sample results of the simulation and some fragments of implementation code.

# CHAPTER 2: MALARIA DYNAMICS AND BASIC MODEL

## 2.1 Introduction

This chapter deals with malaria dynamics and the basic model of our system. In section 2.2, we present a brief background on simulation (Implementation will be a simulation process) and modelling. Section 2.3 describes malaria dynamics, paying attention to the biology of its populations; i.e., human hosts and mosquito vectors. Section 2.4 presents the basic model from which our system is built. This model is based on Ross-MacDonald model, which still is the most used model for much malarial epidimiology (Aron and May [1]).

## 2.2 Background

### *2.2.1 Notion of simulation*

With the development of computer technology, simulation is more and more widely used in many fields of the human society. Simulation techniques not only play very important roles in scientific studies, but also occupy very important places in education, military, entertainment and almost every field that we may imagine.

A simulation is a system that is constructed to work, in some ways, analogously to another system of interest. The constructed system is usually made simpler than the original system so that only the aspects of interest are mirrored. Simulations are commonly used to learn more about the behaviour of the original system, possibly where original system is not available for manipulation. It may not be available because of cost or safety reasons, or it may not have been built yet and one needs to learn about and design a better system. If the purpose of learning is to train novices, then cost, safety, or convenience are likely reasons to work on a simulated system. The simulation may be a computer simulation or a small-scale physical model. An example is a flight simulation, which is widely used in flight training and entertainment. However, a simulator is a computer program, which mimics both the internal behaviour of a real-world system and the input processes that drive or control it. Typically, we simulate rather than experiment with the real world system because:

- The system as yet, does not exist
- Experimentation with the system is expensive, too time consuming, too dangerous

- Experimentation with the system is inappropriate, e.g. disaster planning.

Simulation may not be done when the system is modelled analytically; i.e., when we use logical reasoning to discuss or solve problems. Simulation may be classified in different kinds, which are:

**Monte Carlo Simulation**

Mote Carlo simulations describe stochastic and static processes. That is, systems that do not undergo any modifications over time. A stochastic process is a random function or a collection of random variables, which varies over time. They concern sequences of events governed by probabilistic laws, say (Karlin & Taylor [4]).

**Continuous time simulations**

- Used to model systems that vary continuously over time.
- The systems modelled are dynamic but may be either stochastic or deterministic. A deterministic model is a mathematical model in which the parameters and variables are not subject to random fluctuations, so that the system is at any time defined by its initial conditions.

**Discrete time Simulation**

- Used to model systems which are assumed to change only at a discrete set of points in time (correspond to state changes).
- The systems modelled are dynamic and almost invariably stochastic.

**Combined Discrete/Continuous Simulation (Hybrid)**

- Combination of discrete and continuous variables.

Systems that change with time, called dynamic systems, and/or involve randomness are good candidates for simulation. Modelling complex dynamic systems theoretically need too many simplifications and so the emerging models may not be therefore valid.

Advantages of simulation include:

- In the real life, one may not have enough time to complete the thorough search.
- The ease of making parameter change in order to optimise the performance or test the parameter.

- Compared with mathematical analyses it is possible to construct and study more realistic models.

The main disadvantages are approximation in the model and approximation in the results obtained.

### *2.2.2 Notion of modelling*

A model is defined as a high level specification to abstract from reality a description of a dynamic system. Modelling is a way of thinking and reasoning about systems. There are two main types of models.

- **Physical model**. This kind of model is concrete i.e., it is palpable and visible.

**Example**: Animal and vegetal clones.

- **Mathematical model**. Unlike physical model, this kind of model is abstract. I.e., a series of mathematical demonstrations, which from hypotheses try to prove using logical steps the existence of a system.

**Example**: The simulation an airplane flight which is widely used in flight training and entertainment.

A model can be used to study system behaviour in the design stage, before such systems are built, to communicate a system design and to predict the performance of new systems under varying sets of circumstances.

## 2.3 Malaria dynamics

According to the World Health Organisation (WHO) report [7], malaria parasites are transmitted from one person to another by the female anopheles mosquitoes. The males do not transmit the disease as they feed only on plant juices. There are about 380 species of anopheles mosquitoes, but only 60 or so are able to transmit the parasite. Malaria is caused by protozoan parasites of the genus *Plasmodium*. Four species of *Plasmodium* can produce the disease in humans:

1. *Plasmodium falciparum*
2. *Plasmodium vivax*
3. *Plasmodium ovale*
4. *Plasmodium malaria*

*Plasmodium falciparum* is the most widespread and most dangerous of the four. Untreated it can seriously damage health.

Anopheles breed in water and each of its species has its preferred breeding grounds, feeding patterns and resting place. Plasmodium develops in the gut of the mosquito and is passed on in the saliva of an infectious insect each time it takes a new blood meal. The parasites are then carried by the blood in the victim's liver where they invade its cells and multiply. They are present in the blood and penetrate the red cells, where they multiply again, progressively breaking down the red cells. This induces bouts of fever and anaemia in the infected individual. In cerebral malaria, the infected red cells obstruct the blood vessels in the brain. Other vital organs can also be damaged often leading to the death of the patient. Malaria is diagnosed by the clinical symptoms and microscopic examination of blood. Anti-malarial drugs can normally cure it. The symptoms of fever, shivering, pain in the joints and headache, quickly disappear once the parasite is killed. However in certain regions, the parasites have developed resistance to certain anti-malarial drugs, particularly chloroquine. In endemic regions, where transmission is high, people are continuously infected so that they gradually develop immunity to the disease. Until they have acquired such immunity, children remain highly vulnerable. Pregnant women are also highly vulnerable since natural defence mechanisms are reduced during pregnancy [6].

Mosquitoes bite people as part of their reproduction process. Their movements are motivated by the desire to bite (seek to bite) so, they move towards persons or water when they want to breed. Reproduction is motivated by the environmental conditions (climate) e.g., high temperatures induce the desire to reproduce and so to bite. They may die from assorted causes. As for persons, their reproduction might be motivated by a natural desire or any other reasons. They kill mosquitoes when they become a nuisance because of bites. They die because of malaria or any other causes. They move about carrying out their daily activities or to avoid mosquitoes. Note that, movements of persons might influence the propagation of the malaria. Births and deaths in the population are seen as natural events, although some people can die because of malaria.

The reproduction of malaria germs takes place in human blood or in the salivary glands of the mosquito. The germs are transmitted through mosquito bites. The transmission is done from mosquito to any person and vice versa. I.e., an infectious mosquito may transmit malaria germs to any person through bites and an infectious person may transmit malaria germs to a mosquito when it bites him/her. Malaria germs die because of drugs or any other causes.

## 2.4. The basic model

The basic model from which our model is built is called SEIR model (Susceptible, Exposed, Infectious, Removed). Populations (hosts and vectors) involved in our study are divided into compartments representing disease status. These are compartments referred to as state variables. There is one set of compartments for the human population and another set for the mosquito population. The host and vector populations include the following compartments:

1. **Susceptible**: the fraction of population that is able to contract the disease.
2. **Exposed**: the fraction of population that has contracted the parasites but is not yet able to transmit it.
3. **Infectious**: the fraction of population that develops the disease and may transmit it.
4. **Removed**: the fraction of population that goes out of the system

An infectious mosquito may transmit malaria germs to a susceptible host through a successful bite. If a host acquires malaria germs, he/she becomes exposed. After a certain time, called incubation period, the Exposed host becomes infectious. If a susceptible mosquito bites an infectious host, it may acquire malaria germs; becomes exposed and later infectious after an incubation period. A subject (host or vector) may go out of the system. In this case, it moves from one of the states S, E or I to state R. Hosts may die because of malaria disease or from natural death; that is death due to causes other than malaria. Vectors die only of natural death. Furthermore, each state contains a fraction of the total population. In fact, if $p_S$, $p_E$, $p_I$ and $p_R$ denote respectively the fractional population in each state, at any time $p_S+p_E+p_I+p_R=1$. The fraction of population in each state may vary over time according to the flow of population in the different states. In our study, we present only populations involved in the disease transmission. These include only female anopheles for only them can transmit malaria germs. People of all ages and sexes are concerned since all may be infected.

## 2.5. Concluding remarks

In conclusion, we realise that the application we want to bring to light is a complex structure that demands a well-defined model from which our system would be designed. The choice of the simulation model is justified by the characteristics of the system and targeted objectives. So our model is defined as a Discrete Event simulation model. Because, first, it changes only at discrete points in time which correspond to different states (susceptible, exposed, infectious, removed). Next, the model is stochastic, i.e., governed by probabilistic laws. Once these basic notions and model are defined, a detailed analysis of the model can be done. That will be the main purpose of the next chapter.

# CHAPTER 3: MODEL ANALYSIS

## 3.1 Introduction

This chapter deals with the analysis of our system. We want to quantitatively capture the transmission of malaria. The model is based on stochastic processes (Gihman & Skorhod [2]), the Markov chains (Lipschutz [5]). First, the concepts of stochastic processes and Markov chains are defined in order to understand better our reasoning. Then, the different agents (humans and mosquitoes) defined in the system are described. The different states in the Markov chain model of the system are defined and then their characteristics identified. An interpretation and a description of the Markov chain model obtained are given. Finally, we derive some values, probabilities that will be used for the simulation process.

## 3.2 Basic definitions

- **Stochastic process**

Suppose we observe some characteristics of a system at discrete points in time (labelled 0,1,2…). Let $X_t$ be the value of the system characteristic at time **t**. In most situations, $X_t$ is not known with certainty before time **t** and may be viewed as a random variable. A discrete-time stochastic process is simply a description of the relation among the random variable $X_0, X_1, X_2, \ldots$

- **Finite Markov chain**

A discrete time and discrete space stochastic process $\{X_t\}$ t=1,2,… is a finite Markov chain, if it has the following properties:

i. A finite number of states (M).
ii. The Markovian property:

$$P\{X_{t+1} = j \mid X_1 = k_1, X_2 = i_2, \ldots X_t = i_t\} = P(X_{t+1} = j \mid X_t = i_t) \qquad (1)$$

iii. Stationary transition probabilities. i.e., $P\{X_{t+1} = j \mid X_t = i\} = P\{X_2 = j \mid X_1 = i\}$. A Markov chain is stationary (or time-homogeneous) if the probability law relating to next period's state to the current does not change (remains stationary) over time.

iv. A set of initial probabilities (initial distribution), $P\{X_0 = i\}$ $i = 1,2,\ldots,M$

A finite Markov chain is generally represented by an oriented graph where each node corresponds to one state and each transition probability ($p_{i,j}$) is associated to each oriented edge of the graph. A Markov chain is defined by the square matrix of order M of transition probabilities $P = [p_{i,j}]$.

- **Transition probability**: Probability to move from state **i** to state **j**. **i** may be equal to **j**. The transition probability from state **i** to state **j** is denoted $p_{i,j}$. More generally, the transition probability **n** periods later from state **i** to state **j** is called the n-step transition probability and is denoted $p_{i,j}^{(n)}$.
- **Transition probability matrix**: A transition probability matrix is a square matrix used to display transition probabilities i.e., $P = [p_{i,j}]$. The transition probability matrix n periods later is called the n-step transition probability matrix and is denoted $P^{(n)} = [p_{i,j}^{(n)}]$. It has the following properties:

    i. $0 \le p_{i,j}^{(n)} \le 1$ (i, j, n)=1,2…

    ii. $\sum_{j=1}^{j=M} p_{i,j}^{(n)} = 1$ (i, j, M, n)=1,2…

    iii. $P^{(n)}$ is a square matrix of order M; M is the total number of states of the Markov chain.

- **Probability distribution**: It is a vector that contains probabilities of being in each state of a Markov chain at a given period. The probability of being in any state **i**, n periods later is defined by $\pi_i^{(n)} = P\{x_n = i\}$ $i=1,2,\ldots,M$. The vector $\prod^{(n)}$ is the probability distribution after n periods and is given as: $\prod^{(n)} = [\pi_i^{(n)}]$. The initial probability distribution is the probability of being in each state **i** initially. It is defined as $\pi_i^{(0)} = P\{x_0 = i\} i=1,2,\ldots,M$. The initial distribution is generally supposed to be known.
- **Absorbing state**: If $p_{i,j}$ is the transition probability from state **i** to state **j**, state **j** is said to be absorbing if $p_{i,j}=1$ for **i** = **j**. In other words, a state is absorbing if we cannot leave it once in it; we stay in it forever.
- **Transient state**: A state **i** is transient if there exists a path from **i** to an absorbing state **j**. A path from **i** to **j** is a sequence of transition that begins in **i** and ends in **j**, such that each transition in the sequence has a positive probability of occurring.

- **Steady state**: A state in which a system is not changing. At steady state, the transition probabilities and distribution of probabilities remain constant over time.

### 3.3 Markov chain of our system

Our model is divided into host and vector compartments. We assume that our Markov chain remains stationary. At first, we identify the different states of our system as shown in table 1:

***Table 1. Different states of the system***

| **Compartments** | **Agents concerned** |
|---|---|
| 1 | Births in the population of hosts |
| 2 | Susceptible host |
| 3 | Exposed host |
| 4 | Infectious host |
| 5 | Removed host due to causes other than malaria |
| 6 | Removed host due to malaria |
| 7 | Births in the population of vectors |
| 8 | Susceptible vector |
| 9 | Exposed vector |
| 10 | Infectious vector |
| 11 | Removed vector |

### *3.3.1 Graphical representation of transition probability matrix*

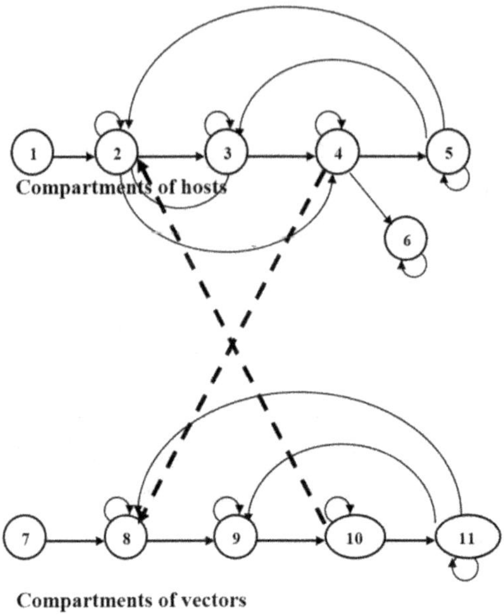


***Figure 1. Graphical representation of transition probability matrix***

### *3.3.2 Transition probability matrix*

Let $P_h$ and $P_v$ be respectively the transition matrix related to host and vector compartments. They are given as:

$$P_h = \begin{array}{c} \\ 1 \\ 2 \\ 3 \\ 4 \\ 5 \\ 6 \end{array} \begin{array}{c} \begin{array}{cccccc} 1 & 2 & 3 & 4 & 5 & 6 \end{array} \\ \begin{bmatrix} 0 & p_{1,2} & 0 & 0 & 0 & 0 \\ 0 & p_{2,2} & p_{2,3} & 0 & p_{2,5} & 0 \\ 0 & p_{3,2} & p_{3,3} & p_{3,4} & p_{3,5} & 0 \\ 0 & p_{4,2} & 0 & p_{4,4} & p_{4,5} & p_{4,6} \\ 0 & 0 & 0 & 0 & p_{5,5} & 0 \\ 0 & 0 & 0 & 0 & 0 & p_{6,6} \end{bmatrix} \end{array}$$

$$P_V = \begin{array}{c|ccccc} & 7 & 8 & 9 & 10 & 11 \\ \hline 7 & 0 & p_{7,8} & 0 & 0 & 0 \\ 8 & 0 & p_{8,8} & p_{8,9} & 0 & p_{8,11} \\ 9 & 0 & 0 & p_{9,9} & p_{9,10} & p_{9,11} \\ 10 & 0 & 0 & 0 & p_{10,10} & p_{10,11} \\ 11 & 0 & 0 & 0 & 0 & p_{11,11} \end{array}$$

### *3.3.3 Interpretation and description of the Markov chain*

Once an agent (host or vector) is infected, the disease develops in its organism and may move from one state to another with a certain probability, as illustrated in figure 1. In our model, we assume that births are into susceptible compartment of host and vector. These births are injected into our system through state 1 for the hosts and state 7 for the vectors.

Thus, in the host compartment model, newborns move initially from state 1 to state 2. The fact that all states except state 5 and 6 are transient states is obvious since, once an agent is born and after a successful bite takes place i.e., a bite which transmits malaria germs, the life cycle of malaria germs (Aron & May [1]) is automatically launched in the host organism and can be stopped by the death or other factors such as malaria drugs or host's immune system. If the disease is stopped by one of these last two factors (malaria drugs or immune system) then the host recovers his/her health and becomes susceptible again. He/she moves from states 3 or 4 to state 2 (transitions: 3 $\rightarrow$ 2 or 4 $\rightarrow$ 2). A host can stay in a given transient state (transition 2 $\rightarrow$ 2, 3 $\rightarrow$ 3 or 4 $\rightarrow$ 4) when there is no change in his/her physiological state. Thus, a host stays in the state 2 until an infectious vector bites a susceptible host, when he/she moves from state 2 to state 3 (transition 2 $\rightarrow$ 3). He/she stays in the state 3 during the incubation period (transition 3 $\rightarrow$ 3). After the incubation period, he/she either develops the disease and then moves from state 3 to state 4 (transition 3 $\rightarrow$ 4) or he/she recovers and goes back to the susceptible state (transition 3 $\rightarrow$ 2). When a host is sick, he/she stays in state 4 (transition 4 $\rightarrow$ 4) and may die because of malaria (transition 4 $\rightarrow$ 6) or recover his/her health (transition 4 $\rightarrow$ 2). Note that, hosts may also die due to causes other than malaria (natural death); this is given by transitions 2 $\rightarrow$ 5, 3 $\rightarrow$ 5 and 4 $\rightarrow$ 5. Once a host enters either state 5 or 6, he/she stays in it forever, since he/she is supposed to be dead.

For vectors, things are slightly different. We assume that vectors do not die because of the malaria disease; i.e., malaria is not considered as a disease for vectors. So, if a vector moves from state **i** to state **j (i > j)**, it cannot come back to state **i**, since it does not recover its health. State 7 represents newborns. Since all newborns are supposed to be susceptible, they initially move from state 7 to state 8 (transition 7 $\rightarrow$ 8). If a vector is susceptible, it is in state 8 and stays there until it bites an infectious host and that bite is infectious. I.e., the bite transmits malaria germs. In this case, it moves from state Susceptible to state Exposed (transition 8 $\rightarrow$ 9) since it already has malaria germs in its blood. Through out the incubation period, it stays in state 9 (transition 9 $\rightarrow$ 9). After the incubation period, it develops the disease, in which case it moves from state Exposed to state Infectious (transition 9 $\rightarrow$ 10). When it is infectious, it stays in this state (transition 10 $\rightarrow$ 10) until death occurs. As for hosts, the death occurs at any moment in any state. This is given by transitions 8 $\rightarrow$ 11, 9 $\rightarrow$ 11 and 10 $\rightarrow$ 11. Dead vectors stay in state 11 forever (transition 11 $\rightarrow$ 11). This death is due to any causes. The matrix $P_v$ is then upper triangular. In both host and vector, we do not consider babies who die at birth.

### 3.4 Study of agents behaviours over time

From the transition probability matrix related to figure 1, we would like to study the behaviour of our system over time. Since we are interested by the host's infection, we are going to seek answers to the following questions in order to quantify the malaria infection:

1. If a host is susceptible (state 2) at time $\mathbf{t_0}$, what is the probability that, **n+1** periods later he/she will be infectious (state 4)?
2. If a vector is susceptible (state7) at time $\mathbf{t_0}$, what is the probability that, at time $\mathbf{t_{n+1}}$ it will be infectious (state 10)?
3. More generally what is the probability that a host or vector initially in state i will be in state **j** at a given time, $\mathbf{t_n}$ ?

These computations will be done at two different levels:

- First, at individual level: At this level, we pay attention to a single agent (host or vector). We would like to capture both in host and vector the following information: the n-step transition probabilities, the probability distribution at each step, the steady state distribution, and the average time spent in the system by an agent. To fulfil these last two objectives, we would need to first define the fundamental matrix denotes N. It is a square matrix that contains only the transition probabilities of transient states in an absorbing Markov chain. If M and s are respectively the total number of states and the number of absorbing states, then N will be a (M-s)(M-s) matrix. The fundamental matrix gives information about the time spent in a system before absorption taking place. In fact starting the process in state **i**, the number of times we expect to visit state **j** before absorption is the (i,j)th entry of N. From a given transient state **i**, the total number of steps expected before absorption is equal to the total number of visits we expect to make to all the transient states **j** ($i \neq j$). This is the sum of all the entries of the i th row of N. In our analysis, the absorption will correspond to death.

- Second, at the population level. This second step allows us to calculate values directly from population data.

### *3.4.1 Individual level*

#### a) Host compartments

**n-step transition probabilities:** Let $P_h^{(1)}$ and $P_h^{(n)}$ be respectively the initial transition probability matrix and the transition probability matrix **n** periods later in the host compartments. A step is defined as a period of time of given length, for instance a day, a week, a month or a year.

For n=1,

$$P_h^1 = \begin{array}{c} \\ 1 \\ 2 \\ 3 \\ 4 \\ 5 \\ 6 \end{array} \begin{array}{c} \begin{array}{cccccc} 1 & 2 & 3 & 4 & 5 & 6 \end{array} \\ \begin{bmatrix} 0 & p_{1,2} & 0 & 0 & 0 & 0 \\ 0 & p_{2,2} & p_{2,3} & 0 & p_{2,5} & 0 \\ 0 & p_{3,2} & p_{3,3} & p_{3,4} & p_{3,5} & 0 \\ 0 & p_{4,2} & 0 & p_{4,4} & p_{4,5} & p_{4,6} \\ 0 & 0 & 0 & 0 & p_{5,5} & 0 \\ 0 & 0 & 0 & 0 & 0 & p_{6,6} \end{bmatrix} \end{array}$$

Let's determine the transition probability matrix **n** periods later

$P_h^{(n)} = P_h^{(1)} P_h^{(n-1)}$ where n > 1 and $P_h^{(k)}$ is the transition matrix k periods later. By recurrence, the (i,j)th element of $P_h^{(n)}$, is defined as a function of the (i,j)th element of $P_h^{(n)}$, as follows:

$$p_{i,j}^{(n)} = \sum_{k=1}^{k=6} p_{i,k}\, p_{k,j}^{(n-1)} \quad (n>1) \text{ and } (i,j) \in \{1, 2, 3, 4, 5, 6\} \tag{2}$$

where $p_{i,k}$ is the (i,k)th the element of $P_h^{(1)}$ and $p_{i,j}^{(n)}$ is the n-step transition probability (the (i,j)th element of $P_h^{(n)}$). Since there are interactions between host and vector compartments through vector bites, the probability $p_{2,3}^{(n)}$, which is directly concerned by these interactions (host-vector), is computed by taking into account the bites. In effect $p_{2,3}^{(n)}$ (transition probability from Susceptible state to Exposed state in host compartment) occurs if an infectious vector bites a susceptible host. These bites also depend on the contact rate between host and vector. Thus, let $H_s$, $V_i$, H, V and $c_{hv}$ be respectively the number of susceptible hosts, the number of infectious vectors, the total number of hosts, the total number of vectors and the contact rate host-vector, assuming that the contact rate is constant over time, we define $p_{2,3}^{(n)}$ as follows:

$$p_{2,3}^{(n)} = \frac{H_s(n-1)}{H(n-1)} * \frac{V_i(n-1)}{V(n-1)} c_{hv} \quad (n > 1) \tag{3}$$

Thus the following general matrix gives the n-step transition probability matrix:

$P_h^{(n)} = [p_{i,j}^{(n)}]\ \forall\ (i,j) \in \{1, 2, 3, 4, 5, 6\}$

**Probability distribution in host compartments:** The goal of this section is to find the probability of a host being in a given state **j** at any time step **n**. Let $\pi_j^{(1)}$ (j = 1 to 6 since we have six states in the host compartments) be the probability that the Markov chain is in state **j** at initial time. Let $\Pi_h^{(1)}$ (column vector) be the initial distribution for the Markov chain related to host compartments, $\Pi_h^{(1)^T} = [\pi_1^{(1)}\ \pi_2^{(1)}\ \pi_3^{(1)}\ \pi_4^{(1)}\ \pi_5^{(1)}\ \pi_6^{(1)}]$. $\Pi_h^{(k)^T}$ denotes the transpose of $\Pi_h^{(k)}$. $\Pi_h^{(1)^T}$ is a vector of probabilities and so all entries must sum to 1 ($\sum_{j=1}^{j=6} \pi_j^{(1)} = 1$). At a given time, we then define the probability vector $\Pi_h^{(n)}$, which is the probability distribution **n** time steps later and is defined as $\Pi_h^{(n)^T} = [\pi_1^{(n)}\ \pi_2^{(n)}\ \pi_3^{(n)}\ \pi_4^{(n)}\ \pi_5^{(n)}\ \pi_6^{(n)}]$ with $\sum_{j=1}^{j=6} \pi_j^{(n)} = 1$. Furthermore, by recurrence, $\Pi_h^{(n)}$ is expressed as a function of $\Pi_h^{(1)}$ and $P_h^{(n)}$ as follows:

$$\Pi_h^{(n)} = \Pi_h^{(1)} P_h^{(n)} \qquad (4)$$

**Fundamental matrix in host compartments**: In general, each element (denoted $n_{i,j}$) of the fundamental matrix is computed using the following formula:

$$n_{i,j} = \begin{cases} 1 + \sum_{k=1}^{M-s} p_{i,k}\, n_{k,j} & \text{if } i = j \\ \sum_{k=1}^{M-s} p_{i,k}\, n_{k,j} & \text{if } i \neq j \end{cases} \qquad (5)$$

where M is the total number of states of the Markov chain, s is the number of absorbing states and M-s is the number of transient states.

Let N be the fundamental matrix, the general form of the fundamental matrix is given by $N = [n_{i,j}]_{(M-s)(M-s)}$. In our specific case (host compartments) M=6, s=2 and $k \in \{1,2,3,4\}$.

**Steady state distribution in host compartments:** Let us study the limit of $\Pi_h^{(n)}$ as n tends to infinity in order to determine the behaviour of the chain in the host compartments after a long time. This limit is called an equilibrium distribution, which is simply the probability distribution at steady state. In fact, after a long run, the probability of being in a given state **i** does not change anymore. This means that after a long run $p_{i,j}^{(n)}=p_{i,j}^{(n-1)}$. In fact, when we take into account newborns, we realise that there will be always agents in the system despite the fact that some of them die. Thus, we would like to determine the number of hosts present in each transient state; i.e., agents which are alive in the system (susceptible, exposed and infectious) at a given step n then find the limit as n tends to infinity.

Let $Z^{(n)^T}$ be a vector (column vector) that contains the number of hosts in each transient state at step n ($Z^{(0)}$ is the initial distribution)

Let $x^T$ be a vector (column vector) that contains the new recruitments (newborns) at the beginning of each step. We would like to find $y=\lim_{n\to\infty} Z^{(n)}$.

$$y=\lim_{n\to\infty} Z^{(n)} \Leftrightarrow y^T=\lim_{n\to\infty} Z^{(n)^T}.$$

Assuming that:

I is an identity matrix of size (M-s)(M-s), Q is a (M-s)(M-s) matrix representing transition probabilities between transient states.

Let us express y independently of $Z^{(n)}$. By recurrence,

$$Z^{(n)T} = Z^{(n-1)T}\, Q + x^T$$

$$\Rightarrow Z^{(1)T} = Z^{(0)T}\, Q + x^T$$

$$\Rightarrow Z^{(2)T} = Z^{(1)T}\, Q + x^T = Z^{(0)T}\, Q^{(2)} + (Q+I)x^T$$

$$\Rightarrow Z^{(n)T} = Z^{(0)T}\, Q^{(n)} + x^T(Q^{(n-1)} + \ldots + Q+I)$$

If $n\to\infty$, $Q^{(n)} \to 0$ and $(Q^{(n-1)} + \ldots + Q+I) \to [I-Q]^{-1}$

and hence $y^T = x^T[I-Q]^{-1} = x^TN$

with $N=[I-Q]^{-1}$. N is the fundamental matrix. In our model, newborns come only from state 1 for host compartments. Let $x^T = [\beta\ 0\ 0\ 0]$ where $\beta$ is the number of newborns at each step in host compartments. $y^T = [\beta\ 0\ 0\ 0]N$. After computation we obtain:

$$\lim_{n\to\infty} Z^{(n)^T} = y^T = \left[\beta \quad (\sum_{k=1}^{4} p_{1,k} n_{k,2})\beta \quad (\sum_{k=1}^{4} p_{1,k} n_{k,3})\beta \quad (\sum_{k=1}^{4} p_{1,k} n_{k,4})\beta\right] \qquad (6)$$

**Interpretation of equation (6):** let $\theta_i$ (i=2,3,4) be the number of hosts in transient state **i** at steady state. Equation (6) means that if we have $\beta$ newborns in host compartments at each time step, at steady state the number of persons in each transient state **i**, is given by:

$$\theta_i = \left(\sum_{k=1}^{4} p_{1,k} n_{k,i}\right)\beta \quad (i=2,3,4) \qquad (7)$$

Recall that i = 2 represents susceptible host, i = 3 represents exposed host and i = 4 represents infectious host. In these 3 states we have the active populations in the system. Finally, the probability of being in state **i** (i=2,3,4) at steady state is given by the following formula:

$$\pi_i^{(\infty)} = \frac{\theta_i}{\sum_{j=2}^{4} \theta_j} \quad (i=2,3,4) \qquad (8)$$

Now, we are going to find the absorbing state distribution at steady state. The absorbing state distribution will give us probabilities to leave the system in each absorbing state. Let $b_{i,j}$ be the probability that a process begins in a transient state **i** and ends up in an absorbing state **j**. $b_{i,j}$ is defined by the following formula: $b_{i,j} = p_{i,j} + \sum_{k=1}^{M-s} p_{i,k} b_{k,j}$

Let B be the matrix that represents transition probabilities from each transient state **i** to each absorbing state **j**. $B = [b_{i,j}]$. In our specific case, **i**$\in\{1, 2, 3, 4\}$ and $j\in\{5, 6\}$

Let z be a vector that contains the initial distribution of transient states and w a vector that contains the absorbing state distribution:

$$w^T = z^T B \qquad (9)$$

with $z^T = [\pi_1^{(1)} \ \ \pi_2^{(1)} \ \ \pi_3^{(1)} \ \ \pi_4^{(1)}]$.

**The average time spent in the system by a host**: If the process begins in a transient state **i**, the average time that will be spent in another transient state **j** is $n_{i,j}$. Recall that $n_{i,j}$ is the (i,j)th element of the fundamental matrix (N).

More generally, if the process begins in a transient state **i**, the average length of time that will be spent in the system is equal to the sum of the ith row of the fundamental matrix. Thus, let $A_i$ be the average length of time spent in the system if the process begins in state **i**:

$$A_i = \sum_{j=1}^{M-s} n_{i,j} \qquad (10)$$

**b) Vector compartments**

In these compartments, the reasoning is the same as the one used in the host compartments. As with the hosts, we would like to determine the n step probabilities, the probability distribution over time, study the behaviour of the chain after a long period of time and determine the average length of time spent in the system by a vector from a given state.

**n-step transition probabilities:** Let $P_V^{(1)}$ and $P_V^{(n)}$ be respectively the initial transition matrix and the transition matrix at step n in the vector compartments,

For n = 1,

$$P_v = \begin{array}{c} \\ 7 \\ 8 \\ 9 \\ 10 \\ 11 \end{array} \begin{array}{c} \begin{array}{ccccc} 7 & 8 & 9 & 10 & 11 \end{array} \\ \begin{bmatrix} 0 & p_{7,8} & 0 & 0 & 0 \\ 0 & p_{8,8} & p_{8,9} & p_{8,10} & p_{8,11} \\ 0 & 0 & p_{9,9} & p_{9,10} & p_{9,11} \\ 0 & 0 & 0 & p_{10,10} & p_{10,11} \\ 0 & 0 & 0 & 0 & p_{11,11} \end{bmatrix} \end{array}$$

The n-step transition probability is defined by the following formula:

$$p_{i,j}^{(n)} = \sum_{k=7}^{k=11} p_{i,k}\, p_{k,j}^{(n-1)} \quad (n>1) \text{ and } (i,j) \in \{7, 8, 9, 10, 11\}$$

Like hosts, vectors acquire infection through infectious bites. In fact, a susceptible vector is infected when it bites an infectious host. In the Markov chain related to vectors, the probability $p_{8,9}$ depends directly on infectious bites. Let $V_s$, $H_i$ and $c_{vh}$ be respectively the number of susceptible vectors, the number of infectious hosts, and the vector-host contact rate, at any step **n**, we define $p_{8,9}^{(n)}$ as follows:

$$p_{8,9}^{(n)} = \frac{V_s(n-1)}{V(n-1)} * \frac{H_i(n-1)}{H(n-1)} c_{vh} \qquad \textit{(11)}$$

Thus the n-step transition probabilities matrix is given by: $P_V^{(n)} = [p_{i,j}^{(n-1)}] \, \forall \, (i,j) \in \{7, 8, 9, 10, 11\}$

**Probability distribution in vector compartments:** Like host compartments, we would like to determine in this section the probability of a vector being in a given state i at any step n. Let $\Pi_V^{(1)}$ (column vector) be the initial probability distribution in the vector compartments. $\Pi_V^{(1)^T} = [\pi_7^{(1)} \ \pi_8^{(1)} \ \pi_9^{(1)} \ \pi_{10}^{(1)} \ \pi_{11}^{(1)}]$. With $\sum_{i=7}^{11} \pi_i^{(1)} = 1$. $\pi_i^{(1)}$ denotes the probability to be in the state i at initial time. At a given time step n, we then define the equilibrium distribution $\Pi_V^{(n)}$, which contains the probabilities of being in the state **i** at **n**. $\Pi_V^{(n)T} = [\pi_7^{(n)} \ \pi_8^{(n)} \ \pi_9^{(n)} \ \pi_{10}^{(n)} \ \pi_{11}^{(n)}]$. With $\sum_{i=7}^{11} \pi_i^{(n)} = 1$. $\Pi_V^{(n)}$ is defined in function of $\Pi_V^{(1)}$ and $P_V^{(n)}$ as follows:

$$\Pi_V^{(n)} - \Pi_V^{(1)} P_V^{(n)} \qquad \textit{(12)}$$

**Fundamental matrix in vector compartments**: Let N be the fundamental matrix. The fundamental matrix will be useful for finding the steady state distribution and the average length time spent in the system by a vector. Note that the horizontal and vertical lines in the matrix below separate the transient and absorbing states

$$P_V = \begin{array}{c} \\ 7 \\ 8 \\ 9 \\ 10 \\ 11 \end{array} \begin{array}{c} \begin{array}{ccccc} 7 & 8 & 9 & 10 & 11 \end{array} \\ \left[\begin{array}{cccc|c} 0 & p_{7,8} & 0 & 0 & 0 \\ 0 & p_{8,8} & p_{8,9} & 0 & p_{8,11} \\ 0 & 0 & p_{9,9} & p_{9,10} & p_{9,11} \\ 0 & 0 & 0 & p_{10,10} & p_{10,11} \\ \hline 0 & 0 & 0 & 0 & p_{11,11} \end{array}\right] \end{array}$$

With M = 5 and s = 1

$$Q=\begin{bmatrix} 0 & p_{7,8} & 0 & 0 \\ 0 & p_{8,8} & p_{8,9} & p_{8,10} \\ 0 & 0 & p_{9,9} & p_{9,10} \\ 0 & 0 & 0 & p_{10,10} \end{bmatrix}_{4\times 4}$$

The fundamental matrix is then derived from the following formula: $N=(I-Q)^{-1}$. Let $n_{i,j}$ the (i,j)th element of the fundamental matrix; if the process begins in a transient state **i** , the average time that will be spent in another transient state j is $n_{i,j}$.

**Steady state distribution in vector compartments**: Let us determine $\Pi_V^{(n)}$ as n tends to infinity in order to capture the behaviour of our system after a long run. As for host compartments, we would like to determine $\lim_{n\to\infty} Z^{(n)}$ where $Z^{(n)}$ is a vector that contains the number of agents found in each transient state at time step n. $y=\lim_{n\to\infty} Z^{(n)} \Rightarrow y^T = x^T N$.

Let $\lambda$ be the number of newborns in vector compartments, which come only from state 7. Thus $x^T = [\lambda\ 0\ 0\ 0] \Rightarrow y^T = [\lambda\ 0\ 0\ 0]\ N$. After computation the number of agents found in each transient state at step n ($y^T$) is:

$$Y^T = \begin{bmatrix} \lambda & \frac{p_{7,8}}{1-p_{8,8}}\lambda & \frac{p_{7,8}\ p_{8,9}}{(1-p_{8,8})(1-p_{9,9})}\lambda & \frac{p_{7,8}\ p_{8,9}\ p_{8,10}}{(1-p_{8,8})(1-p_{9,9})(1-p_{10,10})}\lambda \end{bmatrix} \qquad (13)$$

**Interpretation of equation (13)**: Let $\eta_j$ (j = 8,9,10) be the number of vectors in state **j** at steady state. Equation (13) means that if we have $\lambda$ newborn vectors at each step then at steady state in each transient state **j**, we will have:

$$\begin{cases} \eta_8=\frac{p_{7,8}}{1-p_{8,8}}\lambda \\ \eta_9=\frac{p_{7,8}\ p_{8,9}}{(1-p_{8,8})(1-p_{9,9})}\lambda \\ \eta_{10}=\frac{p_{7,8}\ p_{8,9}\ p_{8,10}}{(1-p_{8,8})(1-p_{9,9})(1-p_{10,10})}\lambda \end{cases} \qquad (14)$$

Recall that **i**=8 represents a susceptible vector, **i**=9 represents an exposed vector and **i** = 10 represents an infectious vector.

Finally, the probability of being in transient state **j** (j = 8, 9, 10) in vector compartments at steady state is given by the following formula:

$$\pi_j^{(\infty)} = \frac{\eta_j}{\sum_{i=8}^{10} \eta_i} \quad (j=8,9,10) \tag{15}$$

As for absorbing state distributions in host compartments, if the initial distribution of transient states is known, the final distribution of absorbing state is defined by $a^T = z^T NR$, where

a is the distribution of absorbing states, z is the initial distribution of transient states, N is the fundamental matrix and R is an (M-s) * s matrix representing transitions from transient states to absorbing states. In our specific case (vector compartments) $R^T = [0 \ p_{8,11} \ p_{9,11} \ p_{10,11}]$

$$a^T = [\pi_7^{(1)} \pi_8^{(1)} \pi_9^{(1)} \pi_{10}^{(1)}] \ NR \tag{16}$$

**The average time spent in the system by a vector**: From the fundamental matrix, we determine the expected number of times that will be spent before absorption takes place. In fact, if we are presently in transient state i, (i = 8, 9, 10) the expected number of times steps that will be spent before absorption takes place is the $(I-Q)_{ij}^{-1}$ element of fundamental matrix, and the average length of time spent in the system from state **i** is:

$$\sum_{j=1}^{M-s} n_{i,j} = \sum_{j=1}^{M-s} (I-Q)_{i,j}^{-1} \tag{17}$$

### *3.4.2 Population level*

In this section, we would like to capture the behaviour of the Markov chain related to our system, at the host and vector population levels. To fulfil our task, we use the census approach (Gihman & Skorhod [3]). In fact, at any step (time), we first determine the population size in each state **i** and then from that population size determine the probability to be in this state **i**. This probability is simply the ratio between the population size in state **i**, and the total population of the system.

**Population size and probabilities over time**: At each step, the population size in each state **i** depends on the incoming flux, the outgoing flux and the population already present in state **i** at the previous time step. The incoming flux is people entering state **i** during a period and the outgoing flux is people leaving state **i** during that period. Thus for a given state **i**, at step **n**:

**Incoming flux** = $\sum_{k\neq i} \alpha_k(n-1)p_{k,i}^{(n-1)} + r_i$ where $\alpha_k(n-1)$ is the population size in state **k** at previous time, $p_{k,i}^{(n-1)}$ is the transition probability from state **k** to state **i** at previous time and $r_i$ is the new recruitments (number of newborns) per unit time in state **i**. Note that $p_{k,i}^{(n-1)}$ is the (k,i)th element of the transition matrix n-1 steps later.

**Outgoing flux** = $\sum_{k\neq i} \alpha_i(n-1)p_{i,k}^{(n-1)}$ where $p_{k,i}^{(n-1)}$ is the transition probability from state **i** to state **k** at previous step.

For any state **i**, the population size is given by the following equation:

$$\alpha_i(n) = \alpha_i(n-1) + r_i + \sum_{k\neq i} \alpha_k(n-1)p_{k,i}^{(n-1)} - \sum_{k\neq i} \alpha_i(n-1)p_{i,k}^{(n-1)} \qquad \textit{(18)}$$

The total population of the whole system at step **n** is given by $\sum_{i=1}^{M-s} \alpha_i(n)$. Recall that M-s is the number of transient states of our Markov chain. Let $\varepsilon_i^{(n)}$ be the probability to be in state **i** at step **n** it is given as:

$$\varepsilon_i^{(n)} = \frac{\alpha_i(n-1)}{\sum_{k=1}^{M-s} \alpha_k(n-1)} \qquad \textit{(19)}$$

Let $\varepsilon_i^{(1)}$, $\alpha(1)$ and $\alpha_i(1)$ respectively denote the probability of being in state **i**, the total population size and the population size in state **i** at initial time. Then $\alpha_i(1)= \varepsilon_i^{(1)}$ $\alpha(1)$. Note that $\varepsilon_i^{(1)}$ and $\alpha(1)$ are known (initial values).

In our model, the new recruitments occur only into the susceptible states. These are from state 1 for the host and state 7 for the vector. Thus, $r_i$=0 for i = 1 i = 7.

**Steady-state census**: For a long-term planning purpose, we would like to predict the steady state, i.e., the population size of each state in their respective compartment at infinity. From this population size we compute the probability to be in each state **i**. Note that at this stage, we are more interested in the population alive in the process; i.e., in the states susceptible, exposed and infectious. Let us determine the number of hosts and vectors present in those states and the probabilities to be in each of those states at infinity.

More formally, consider a system whose agents are classified at any point in time into one of the transient states, susceptible, exposed and infectious. At any number of time steps, a fraction of agents who begin the process in the transient state **i** at time **t**, begins the next period in transient state **j**. while another fraction leaves completely the system by moving to an absorbing state. This last fraction is considered as lost. Recall that in our model, people present in an absorbing state are dead.

Thus, let $\varphi_i(t)$ be the number of agents in each transient state **i** at the beginning of each period.

A question of natural interest is whether $\varphi_i(t)$ approaches a limit as t grows large. This limit is found by computing the steady state census (Gihman & Skorhod [3]).

Let $\varphi$ be the steady state census, which contains the limit size of the population in each transient state **i**. Then $\varphi=(\varphi_1, \varphi_2,\ldots, \varphi_{M-s})$, where M-s is the number of transient states of the Markov chain studied. The steady state census is found by solving a system of M-s equations that is derived as follows:

Number of people entering state **i** during each period = $r_i + \sum_{k\neq i} \varphi_k p_{k,i}$

Number of people leaving state **i** during each period = $\varphi_i \sum_{k\neq i} p_{i,k}$

$\sum_{k\neq i} p_{i,k} = 1-p_{i,i}$, hence $\varphi_i \sum_{k\neq i} p_{i,k} = \varphi_i(1-p_{i,i})$. Then the equation used to compute the steady-state census is given by:

$$r_i + \sum_{k\neq i} \varphi_k p_{k,i} = \varphi_i(1-p_{i,i}).\ (i= 1,2\ldots\ M\text{-}s) \qquad \textit{(20)}$$

$p_{k,i}$ is the transition probability from state **k** to state **i** as mentioned in our Markov chain. Let us determine the steady state census and the probability to be Susceptible, Exposed or Infectious at infinity. By developing equation (20) for host compartments we have:

$$\begin{cases} r_1 + \varphi_2 p_{2,1} + \varphi_3 p_{3,1} + \varphi_4 p_{4,1} = \varphi_1(1-p_{1,1}) \\ r_2 + \varphi_1 p_{1,2} + \varphi_3 p_{3,2} + \varphi_4 p_{4,2} = \varphi_2(1-p_{2,2}) \\ r_3 + \varphi_1 p_{1,3} + \varphi_2 p_{2,3} + \varphi_4 p_{4,3} = \varphi_3(1-p_{3,3}) \\ r_4 + \varphi_1 p_{1,4} + \varphi_2 p_{2,4} + \varphi_3 p_{3,4} = \varphi_4(1-p_{4,4}) \end{cases} \qquad (21)$$

$r_2 = r_3 = r_4 = 0$ since the recruitments (births) occur only in state 1. Also, in our Markov chain $p_{3,1} = p_{4,1} = p_{1,3} = p_{4,3} = p_{1,4} = p_{2,4} = 0$. From the system of equation above, it follows that:

$$\varphi_1 = r_1 \qquad \varphi_2 = \frac{(1-p_{3,3})(1-p_{4,4})p_{1,2}}{p_{2,3}\mu} r_1 \quad \varphi_3 = \frac{(1-p_{4,4})p_{1,2}}{\mu} r_1 \quad \varphi_4 = \frac{p_{1,2}p_{3,4}}{\mu} r_1$$

with $\mu = \frac{(1-p_{3,3})(1-p_{4,4})}{p_{2,3}p_{3,4}} + \frac{(1-p_{3,3})p_{3,2} + p_{3,4}p_{4,2}}{(1-p_{2,2})p_{3,4}}$.

Thus if there are $r_1$ recruitments the populations in states 2, 3 and 4, would be respectively $\varphi_2$, $\varphi_3$ and $\varphi_4$ at steady state. The steady state census is given by:

$$(\varphi_2 \ \varphi_3 \ \varphi_4) = \left( \frac{(1-p_{3,3})(1-p_{4,4})p_{1,2}}{p_{2,3}\mu} r_1 \quad \frac{(1-p_{4,4})p_{1,2}}{\mu} r_1 \quad \frac{p_{1,2}p_{3,4}}{\mu} r_1 \right) \qquad (22)$$

Let $\varepsilon_i^{(\infty)} = \frac{\varphi_i}{\sum_{k=2}^{M-s} \varphi_k}$ be the probability to be in state **i** as we tend to infinity. Then:

$$\begin{cases} \varepsilon_2^{(\infty)} = \frac{\varphi_2}{\varphi_2 + \varphi_3 + \varphi_4} \\ \varepsilon_3^{(\infty)} = \frac{\varphi_3}{\varphi_2 + \varphi_3 + \varphi_4} \\ \varepsilon_4^{(\infty)} = \frac{\varphi_4}{\varphi_2 + \varphi_3 + \varphi_4} \end{cases} \qquad (23)$$

and by developing equation (20) in vector compartments we have:

$$\begin{cases} r_7+\varphi_8 p_{8,7}+\varphi_9 p_{9,7}+\varphi_{10} p_{10,7}=\varphi_7(1-p_{7,7}) \\ r_8+\varphi_7 p_{7,8}+\varphi_9 p_{9,8}+\varphi_{10} p_{10,8}=\varphi_8(1-p_{8,8}) \\ r_9+\varphi_7 p_{7,9}+\varphi_8 p_{8,9}+\varphi_{10} p_{10,9}=\varphi_9(1-p_{9,9}) \\ r_{10}+\varphi_7 p_{7,10}+\varphi_8 p_{8,10}+\varphi_9 p_{9,10}=\varphi_{10}(1-p_{10,10}) \end{cases} \tag{24}$$

As for host compartments $r_8 = r_9 = r_{10} = 0$ since recruitments occur only in state 7. According to our Markov chain for vector compartments $p_{8,7} = p_{9,7} = p_{10,7} = p_{10,8} = p_{10,9} = p_{7,10} = p_{8,10} = 0$. After solving the system of equation (24) we have:

$$\varphi_7=r_7 \qquad \varphi_8=\frac{p_{7,8}}{1-p_{8,8}}r_7 \qquad \varphi_9=\frac{p_{7,8}p_{8,9}}{(1-p_{8,8})(1-p_{9,9})}r_7 \qquad \varphi_{10}=\frac{p_{7,8}p_{8,9}p_{9,10}}{(1-p_{8,8})(1-p_{9,9})(1-p_{10,10})}r_7$$

$$(\varphi_8\ \varphi_9\ \varphi_{10})=\left(\frac{p_{7,8}}{1-p_{8,8}}r_7 \quad \frac{p_{7,8}p_{8,,9}}{(1-p_{8,8})(1-p_{9,9})}r_7\right) \frac{p_{7,8}p_{8,9}p_{9,10}}{(1-p_{8,8})(1-p_{9,9})(1-p_{10,10})}r_7 \tag{25}$$

As for the host compartments, if there are $r_7$ recruitments (births), the population in states 8, 9 and 10, would be respectively $\varphi_8$, $\varphi_9$ and $\varphi_{10}$ at steady state. Hence, the probabilities to be respectively Susceptible, Exposed and Infectious at infinity are:

$$\begin{cases} \varepsilon_8^{(\infty)}=\dfrac{\varphi_8}{\varphi_8+\varphi_9+\varphi_{10}} \\ \varepsilon_9^{(\infty)}=\dfrac{\varphi_9}{\varphi_8+\varphi_9+\varphi_{10}} \\ \varepsilon_{10}^{(\infty)}=\dfrac{\varphi_{10}}{\varphi_8+\varphi_9+\varphi_{10}} \end{cases} \tag{26}$$

### 3.5 Concluding remarks

From our analyses, we conclude that at individual level, every agent, which enters the system will end an absorbing state where it stays forever. At population level, things are quite different. Despite the fact that some agents leave the system, population is always maintained alive through new recruitments. But, if there are no new births, population might disappear with time.

# CHAPTER 4: IMPLEMENTATION, ANALYSIS OF RESULTS AND VALIDATION

## 4.1 Introduction

The program ***Sipa*** is a simulation of the malaria that allows us to observe values and their variations over time. In this chapter, we first present ***Sipa*** and its simulation runs. Next we validate the results of simulation runs. This validation should be normally made against empirical data. But unfortunately, the data available to us from biological surveys (Titanji et al. [12], Wondji et al. [13]) were not compatible with the model presented in chapter 3. Nonetheless, we formulate a deterministic model to describe the same scenario as the stochastic model described in chapter 3. A numerical simulation of it is performed followed by a statistical correlation (Spiegel [7]) of it results with those from the stochastic model. Statistical tests help decide whether or not the correlation is good. Since we do not have any clinical or appropriate data, the simulations were initialised with reasonable empirical values and expect a good correlation of their outputs. The results are discussed in section 4.5.

## 4.2 Simulation run

***Sipa*** is built with the programming language C++ builder which is an oriented object programming language. The implementation is divided into two main parts: the stochastic and the deterministic simulation. Before any execution, the user must first choose the kind of simulation he/she wants to process. Initial parameters are input through the Graphic User Interface (GUI) built for this effect. In what follows GUI commands and expressions are given in bold (e.g. **Close**). The main GUI (figure 2) is the screen initially displayed once ***Sipa*** is launched. It allows the user to choose either the stochastic or deterministic simulation to run.

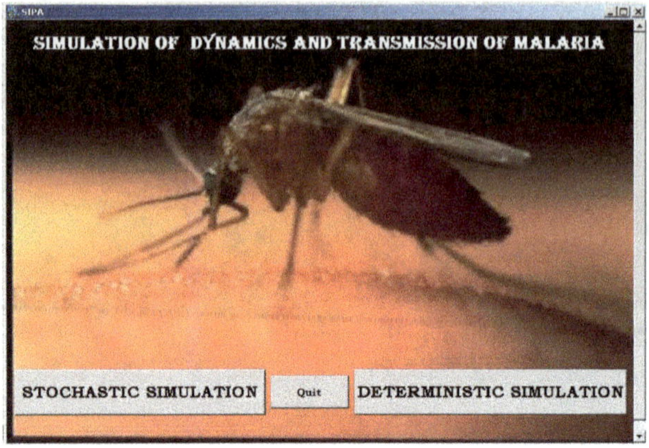


*Figure 2. Main GUI of Sipa*

The command **stochastic Simulation** runs the stochastic simulation and **deterministic Simulation** runs the deterministic. **Quit** exits ***Sipa***. If **stochastic Simulation** is executed the screen showed in figure 3 is displayed. Initial simulation parameters are then set. If **deterministic model** is executed, the screen of figure 4 is used instead.

### *4.2.1 GUI parameter setting*

For the stochastic simulation we have:

- Top and left **Transition (Host):** Initial transition probability matrix of hosts.
- Top and far right **Transition (vector):** Initial transition probability matrix of vectors.
- For the both matrices, some transition probabilities are automatically initialised/defined by the system since they never change over time. This is the case for transition probabilities from each absorbing state (Removed) to any other state (cf. figure 1, chapter 3).
- Middle left **Distribution (host)**: Initial distribution of probability of hosts
- Middle right **Distribution (vector)**: Initial distribution of probability of vectors.
- For any simulation, we always assume that initially, there are no newborns and deaths (removed) in the system. Hence, the probabilities of these compartments are initialised to zero.
- Bottom left: Initial rates (contact, births and deaths).

- Bottom right: Initial host and vector population sizes, time step and total duration of the simulation
- Probabilities are real numbers in the range 0 to 1 inclusive. Population sizes and time steps are positive integers. A suitable warning message is issued for erroneous input and the simulation automatically stopped. The time step is expressed in days and the duration of the simulation in years. For instance, if the time step is a week or a month, the user inputs 7 or 30 respectively. If the duration input is 10, the simulation runs for ten years.

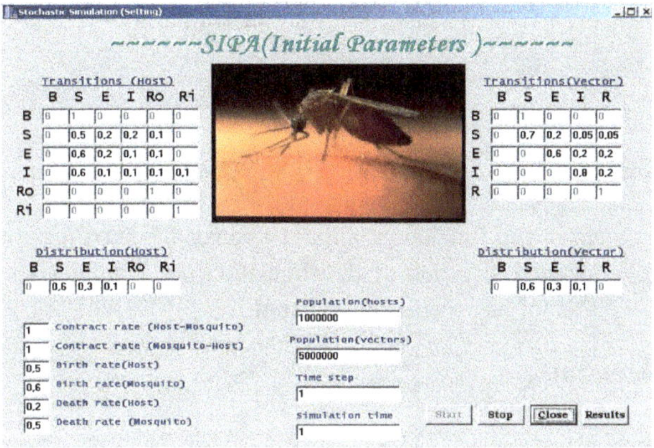


*Figure 3. GUI for the parameters of the stochastic simulation.*

The command **Start** starts the simulation, **Stop** stops the simulation, **Close** allows closing ***Sipa*** and **Results** opens a new window and shows outputs. **Stop** and **Results** are enabled if and only if **Start** is executed.

For the deterministic simulation we have:

- Top left **host compartments**: Initial values of hosts.
- Top right **vectors compartments**: Initial values of vectors.

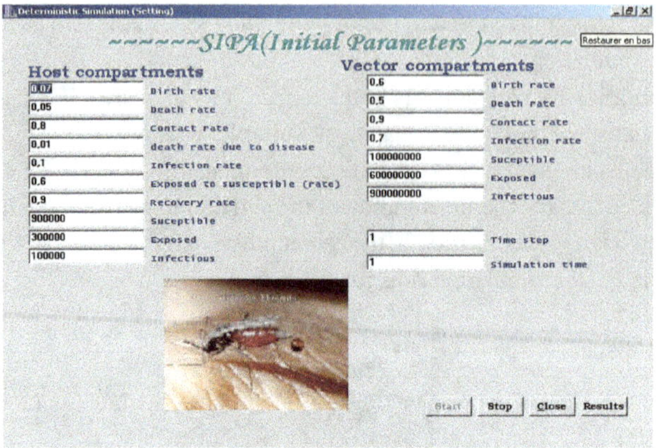


***Figure 4. GUI for the initial parameters of the deterministic simulation.***

In the deterministic simulation, all initial values for hosts and vectors are real numbers. The time step, duration of the simulation and commands have the same interpretation/role as for the stochastic simulation.

### *4.2.2 GUI for output*

For the stochastic simulation we have:

- Top left **Transition (Host):** n-step transition probability matrix of hosts.
- Top right **Transition (Vector):** n-step transition probability matrix of vectors.
- Middle left **Distribution (hosts):** distribution of probability of hosts at time t.
- Middle right **Distribution (vector):** distribution of probability of vectors at time t.
- Bottom left: population sizes in host compartments.
- Bottom near left: population sizes in vector compartments.
- Bottom centre: **Time**: remaining time of simulation expressed in days. Initially, **Time** contains the total simulation time expressed in days. After each time step, this value is decremented by the value of step. It is zero at the end of simulation.

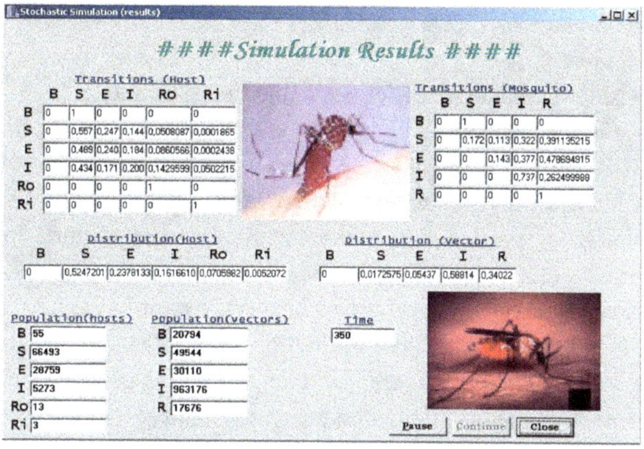


*Figure 5. GUI for output results of the stochastic model.*

The commands **Pause** stops the simulation at a given time t, **Continue** continues the simulation after a pause. **Close** allows closing the windows and going back to figure 3. **Continue** is enabled if and only if the command **Pause** is executed.

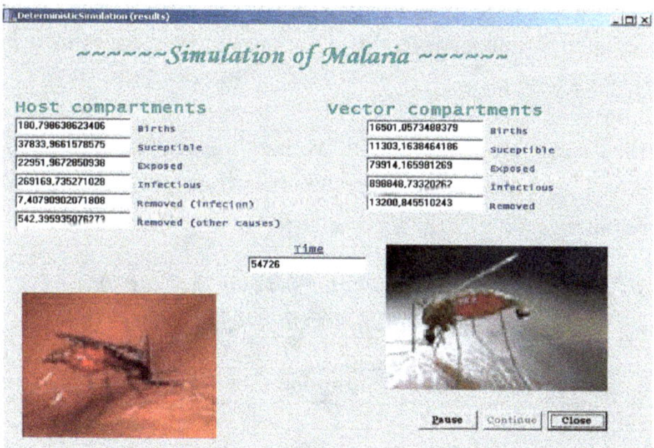


*Figure 6. GUI for output results of the deterministic simulation.*

The results generated during a simulation run are simultaneously displayed on screen and sent into output files (for further analyses) that are created automatically at run time. See appendixes A and B respectively for the algorithms and output data.

## 4.3 Deterministic model

We consider the same states defined in the stochastic model i.e, S, E, I and R. At each time t, for the population of hosts, there are $S_h$ susceptible, $E_h$ Exposed, $I_h$ infectious, $R_h$ Removed, $B_h$ newborns, $R_i$ hosts removed due to infected and $R_o$ hosts removed due to causes other than the infection. For the population of vectors, we have $S_v$ susceptible, $E_v$ exposed, $I_v$ Infectious and $R_v$ removed vectors. Removed hosts include those who die of disease or other causes i.e., $R_h = R_i + R_o$. The subscript h denotes the hosts and v the vectors. We assume that the births are into the susceptible populations of hosts and vectors. The population sizes are not constant over time. The birth rate for the hosts and vector are $\lambda_h > 0$ and $\lambda_v > 0$ respectively. The susceptible hosts and vectors become infectious with rate $\nu_h > 0$ and $\nu_v > 0$ respectively. Infectious host recovers health i.e., move into the susceptible state with rate $r_h$. The per capita death rate for hosts and vectors are respectively $\mu_h$ and $\mu_v$. Natural deaths (that are deaths due to causes other than malaria) occur in the susceptible, exposed and infectious compartments. Death due to malaria occurs only in the infectious compartment of hosts. We also assume that mosquitoes do die only from natural causes. Infectious hosts die from the disease at the additional rate $\gamma_h > 0$ and may transfer infection to a susceptible vector with contact rate $c_{hv}$, and with a contact rate $c_{vh}$, infectious vectors transfer infection to a susceptible host, an exposed host may become susceptible with rate$\phi_h > 0$.

From the description given above, we may then define the equation (transition rate) of each basic event and the corresponding transitions relating to them in the stochastic model.

***Table 2. Basic equations for the elementary events that constitute the deterministic version of the stochastic model.***

| Basic event | Transition rate | Transition |
|---|---|---|
| *In the host population* | | |
| Birth | $\lambda_h N_h$ | 1→2 |
| Natural death -susceptible | $\mu_h(N_h)S_h$ | 2→5 |
| Exposure to infection | $C_{vh}I_vS_h / N_h$ | 2→3 |
| Emergence of a new infective | $\nu_h E_h$ | 3→4 |
| Natural death -Exposed | $\mu_h(N_h)E_h$ | 3→5 |
| Recovery- ill and infective | $r_h I_h$ | 4→2 |
| Natural death –infectious | $\mu_h(N_h)I_h$ | 4→5 |
| Disease induced death | $\gamma_h I_h$ | 4→6 |
| Exposed becoming susceptible | $\phi_h E_h$ | 3→2 |

| *In the vector population* | | |
|---|---|---|
| Birth | $\lambda_v N_v$ | 7→8 |
| Natural death -susceptible | $\mu_v(N_v)S_v$ | 8→11 |
| Exposure to infection | $C_{hv}I_hS_v / N_h$ | 8→9 |
| Natural death -Exposed | $\mu_v(N_v)E_v$ | 9→11 |
| Emergence of a new infective | $\nu_v E_v$ | 9→10 |
| Natural death –infectious | $\mu_v(N_v)I_v$ | 10→11 |

The notation i→j means "move from state i to state j". Each transition from state i to j is characterised by a transition probability value pij (see Table 1 and Figure 1 in chapter 3).

For the deterministic model, flows from compartment are described by ordinary differential equations based on standard mass action, i.e., the rate of change of a class is directly proportional to the density of the state variables in the given class (c.f. Teboh [11]). These equations are represented by the general form $\frac{dx}{dt}=ax+b$ where $a$ is the fractional rate which may be constant or dependent on other state variables, $b$ is a removal and/or recruitment term, and $x$ is a state variable. For both populations (host and vector) the contact rate between infectious and susceptible class of the other determine flows from the susceptible to the exposed compartment. Flows from each of the remaining compartments to the next are assumed to depend on the average residence times for each compartment. Assuming that $N_h$ is the total population of hosts at time t the effective active population is $N_h = S_h + E_h + I_h$. Thus, in the absence of disease, $N_h = S_h$. So we model the susceptible in the absence of disease by the equation $\frac{dS_h}{dt}=(\lambda_h-\mu_h(N_h))S_h$ with $S_h(0) = S_h^0$. We write $\mu_h$ as $\mu_h(N_h)$ instead of $\mu_h(S_h)$ because natural death depends on the total population and not only on the susceptible population. For the vectors, the active population at time t is $N_v = S_v + E_v + I_v$. In the absence of parasites, $N_v = S_v$ and the susceptible population of vectors is modelled by the equation $\frac{dS_v}{dt}=(\lambda_v-\mu_v(N_v))S_v$ with $S_v(0) = S_v^0$. However, in the presence of the disease and the parasites in host and vector population respectively, using the basic equations defined in table 2, the susceptible, exposed, infectious and removed hosts due to disease should satisfy the following equations.

$$\left.\begin{aligned}
\frac{dS_h}{dt} &= \lambda_h N_h + r_h I_h + E_h \phi_h - (\mu_h(N_h) + \frac{c_{vh} I_v}{N_h}) S_h \\
\frac{dE_h}{dt} &= \frac{c_{vh} I_v S_h}{N_h} - (\mu_h(N_h) + \nu_h + \phi_h) E_h \\
\frac{dI_h}{dt} &= \nu_h E_h - (\mu_h(N_h) + r_h + \gamma_h) I_h \\
\frac{dR_i}{dt} &= \gamma_h I_h \\
\frac{dN_h}{dt} &= N_h + (\lambda_h - \mu_h(N_h))\, N_h - \gamma_h I_h \\
\frac{dS_v}{dt} &= \lambda_v(N_v) N_v - (\mu_v(N_v) + \frac{c_{hv} I_h}{N_h}) S_v \\
\frac{dE_v}{dt} &= \frac{c_{hv} I_h S_v}{N_h} - (\mu_v(N_v) + \nu_v) E_v \\
\frac{dI_v}{dt} &= \nu_v E_v - \mu_v(N_v) I_v
\end{aligned}\right\} \qquad \textbf{\textit{(27 a)}}$$

Initially, $S_h(0) = S_h^0$ $E_h(0) = E_h^0$ $I_h(0) = I_h^0$ $S_v(0) = S_v^0$ $E_v(0) = E_v^0$ $I_v(0) = I_v^0$

$R_i(0) = 0$ $N_h^0 = S_h^0 + E_h^0 + I_h^0$

The events that do not directly depend on malaria i.e., births and natural deaths are modelled by the following equations

$$\left.\begin{aligned}
\frac{dB_h}{dt} &= \lambda_h N_h \\
\frac{dR_o}{dt} &= \mu_h(N_h) N_h \\
\frac{dB_v}{dt} &= \lambda_v N_v \\
\frac{dN_v}{dt} &= (\lambda_v - \mu_v(N_v)) N_v \\
\frac{dR_v}{dt} &= \mu_v(N_v) N_v
\end{aligned}\right\} \qquad \textbf{\textit{(27 b)}}$$

*Initially,* $B_h(0) = B_v(0) = R_o(0) = R_v(0) = 0 \quad N_V^0 = S_V^0 + E_V^0 + I_V^0$

Equations (27 a) and (27 b) are those implemented into ***Sipa*** as deterministic model.

## 4.4 Analysis of results

Different forms of initial data were used to simulate both models (stochastic and deterministic). The results found may be divided in two main groups. For the both models, if $\mu_v > \lambda_v$ and $\mu_h > \lambda_h$, whatever the values of the others parameters, that were all positive, the population sizes of hosts and vectors decrease progressively over time from the initial values we started with and tend to zero. Second, if at initial time $S_v > 0$ and/or $S_h > 0$, $I_h \geq 1$ and $I_v \geq 1$, $E_h \geq 0$ and/or $E_v \geq 0$, $\mu_v < \lambda_v$ and $\mu_h < \lambda_h$,, whatever the values of the other parameters that were all positive, the results obtained were qualitatively the same for all experiments after a long time. The susceptible growth up and then reach a steady state. The exposed and infectious growth downs until reach their respective steady state.

In the graphs presented below, the susceptible, exposed, infectious, newborns and removed are represented respectively by their fractional values i.e., the population in a compartment over the total compartment. E.g., for susceptible hosts this is $S_h/(S_h + E_h + I_h + B_h + R_i + R_o)$. Note that labels suffixed with the letter "M" refer to Markov chain model and those followed by the letter "D" refer to deterministic model. Initial values used and output results are given in appendix B.

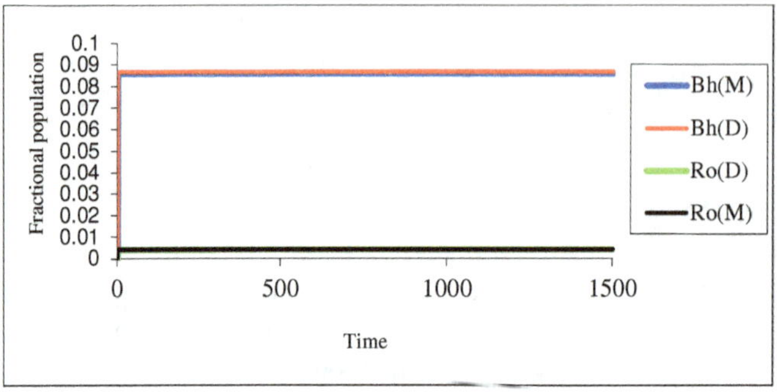


***Figure 7. Long-term behaviour of newborns and removed hosts due to causes other than malaria.***

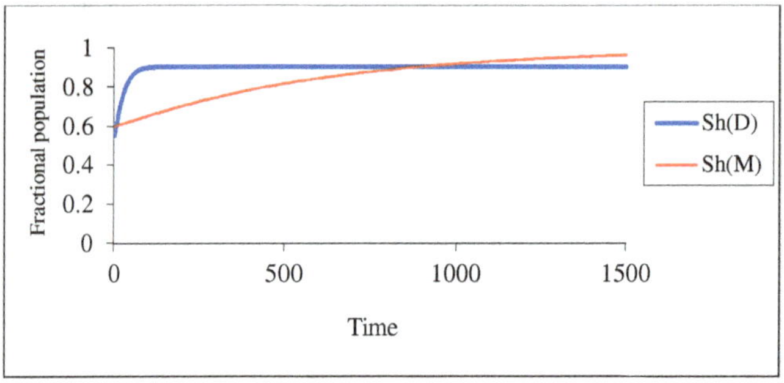


***Figure 8. Long-term behaviour of susceptible hosts.***

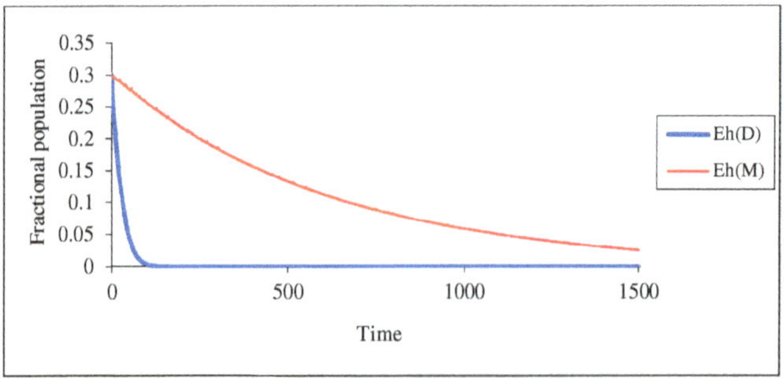


***Figure 9. Long-term behaviour of exposed hosts.***

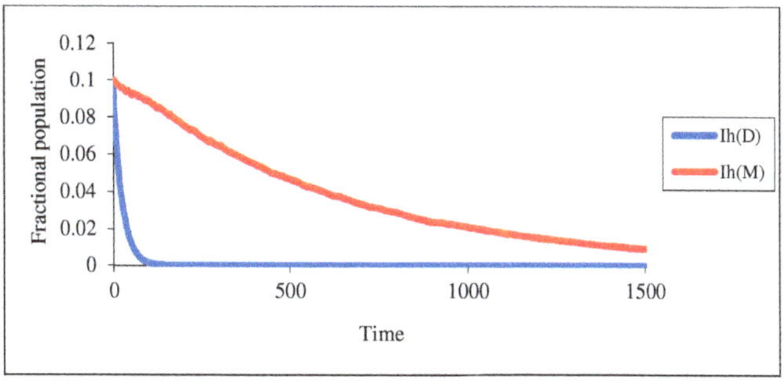


***Figure 10. Long-term behaviour of infectious hosts.***

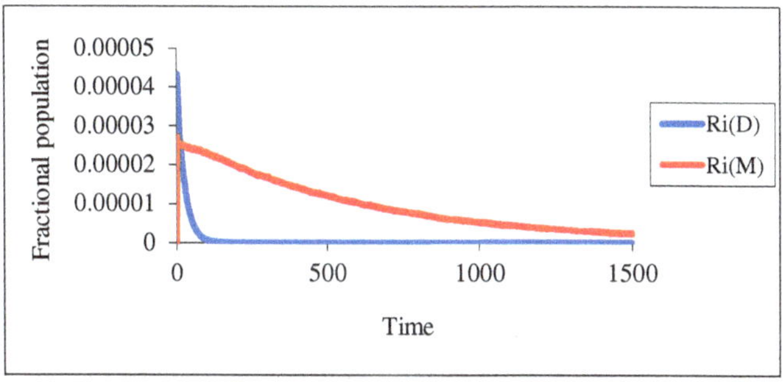


*Figure 11. Long-term behaviour of removed hosts due to malaria.*

For the vectors, we realised that the curves of susceptible rise exponentially from the initial values we started with and then reached a steady state. At the same time, the curves of exposed and infectious decrease continuously to their respective steady state.

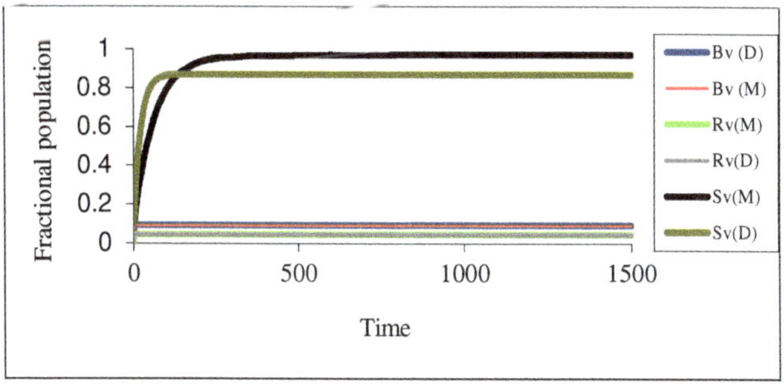


***Figure 12. Long-term behaviour of newborns, removed and susceptible vectors.***

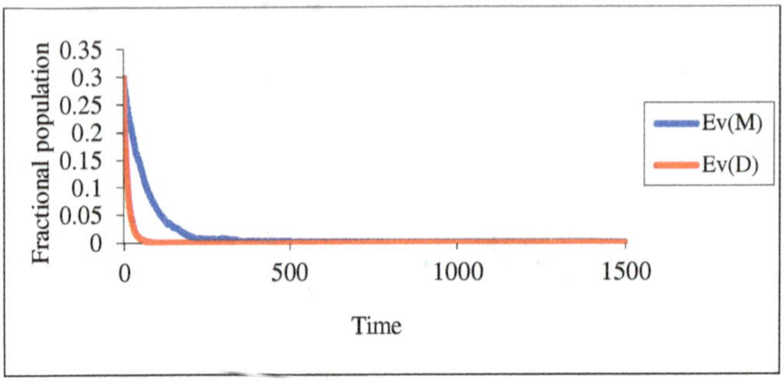


***Figure 13. Long-term behaviour of exposed vectors.***

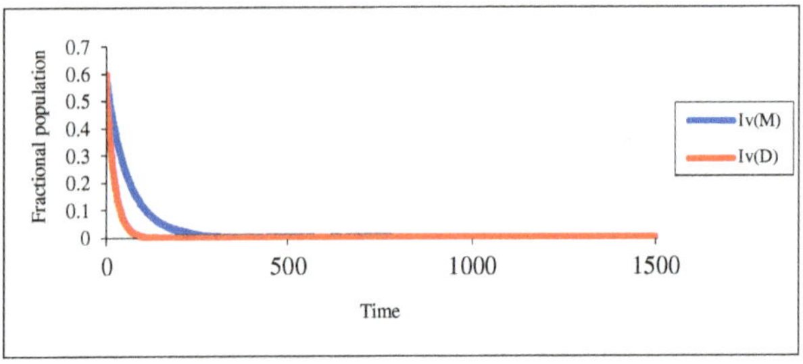


***Figure 14. Long-term behaviour of infectious vectors.***

### *4.4.1 Correlation and hypothesis test analysis*

Correlation is a relationship between two statistical variables x and y. If we have a set of n observations $(x_1, y_1), (x_2,y_2),\ldots,(x_n,y_n)$, the correlation gives the relationship between all points $(x_i,y_i)$. For $1 \le i \le n$ and $x_i$ belong to x and $y_i$ belong to y. To achieve this goal we use the Spearman Rank Correlation Coefficient denoted r (Spiegel [7]).

$$r=1-\frac{6\sum_{i=1}^{n} d_i^2}{n(n^2-1)} \tag{28}$$

$d_i$ is the difference (deviation) between $x_i$ and $y_i$ and n is the number of pairs $(x_i,y_i)$. r must respect the following condition: $-1 \le r \le 1$. $x_i$ could represent the susceptible hosts obtained from the stochastic simulation and $y_i$ those obtained from the deterministic model at the same time step i. The closer r is to either +1 or -1, the stronger the correlation. If the correlation is 0 or very close to 0, there is no association between the two populations.

Assuming that $\rho$ and r are respectively the correlation coefficient of two populations and the correlation coefficient of a sample of those populations, the estimation of $\rho$ can be done as function of r. According to Spiegel [7],

- If $\rho = 0$ is considered as the null hypothesis, one may use the statistics having a student distribution (also called t distribution ) because r is supposed to have a symmetrical distribution that is ., $t_{(1-\alpha)} = -t_\alpha$. Where $\alpha$ and $1-\alpha$ are respectively the significance and the confidence levels of the test. Thus, if $\rho = 0$, $t=r\sqrt{\frac{n-2}{1-r^2}}$ has a student distribution at n-2 degree of freedom. Thus, if r tends to $\pm 1$ then t tends to infinity, which implies a very good correlation.
- If $\rho \neq 0$ is considered as the null hypothesis, we use the fact that the statistics $Z=\frac{1}{2}\ln\frac{1+r}{1-r}$ (also called Z-transformation of Fisher) is approximately normally distributed with a mean $\mu_Z=\frac{1}{2}\ln\left(\frac{1+\rho}{1-\rho}\right)$, a standard deviation $\sigma_Z=\frac{1}{\sqrt{n-3}}$ and a reduced variable $z=\frac{Z-\mu_Z}{\sigma_Z}$

Let $H_0$ and $H_1$ be respectively the null and the alternative hypotheses. For two-sided tests, if $-t_{(1-\alpha)} \leq t \leq t_{(1-\alpha)}$ or $-z_{(1-\alpha)} \leq z \leq z_{(1-\alpha)}$, $H_0$ cannot be refuted else $H_1$ is accepted. Unfortunately, $\rho$ is unknown and so $\mu_z$ and z cannot be computed since they depend on $\rho$. We will use the student distribution since t is only a function of r, which can be computed from data values. The hypothesis test is formulated as follows:

**$H_0$**: $\rho = 0$ There is no relationship between populations.

**$H_1$**: $\rho \neq 0$ There is relationship between populations.

Recall that we have two kinds of populations that are deterministic and stochastic.

Let the significance level $\alpha=0.01$ and our number of observations be the total time of simulation (n = 1500 months). For the initial values used, that duration was long enough to have a steady state. According to student law (Spiegel [7]) , if n>120 then the number of degree of freedom is infinity; as a result, $t_{0.99}=2.576$. Thus, if $-2.576 \leq t \leq 2.576$ then $H_0$ cannot be refuted.

### *4.4.2 Results of the correlation and of hypothesis tests*

***Table 3. Correlation coefficient and results of student t test on the stochastic and deterministic data.***

| Variable (population) | Correlation coefficient (r) | Student distribution (t) |
|---|---|---|
| $B_h$ (stochastic model)<br>$B_h$(deterministic model) | 0.99999999 | 3703199.57 |
| $S_h$ (stochastic model)<br>$S_h$ (deterministic model) | 0.42650584 | 18.5527882 |
| $E_h$ (stochastic model)<br>$E_h$ (deterministic model) | 0.43119762 | 18.8031811 |
| $I_h$ (stochastic model)<br>$I_h$ (deterministic model) | 0.39960848 | 17.1514015 |
| $R_i$ (stochastic model)<br>$R_i$ (deterministic model) | 0.40251332 | 17.3000806 |
| $R_o$ (stochastic model)<br>$R_o$ (deterministic model) | 0.99999982 | 66392.8179 |
| $B_v$ (stochastic model)<br>$B_v$ (deterministic model) | 0.99999999 | 3703199.57 |
| $S_v$ (stochastic model)<br>$S_v$ (deterministic model) | 0.86545319 | 67.967327 |
| $E_v$ (stochastic model)<br>$E_v$ (deterministic model) | 0.74773484 | 44.306423 |
| $I_v$ (stochastic model)<br>$I_v$ (deterministic model) | 0.89068182 | 77.0819423 |
| $R_v$ (stochastic model)<br>$R_v$ (deterministic model) | 0.99999999 | 3703199.57 |

The strength of relationship between data is fairly good since r approaches 1 or is equal to 1. For hypothesis test, H1 cannot be refuted since $t > t0.99$, for any r.

## 4.5 Discussion

We have defined a model of the dynamics and transmission of malaria using Markov chains. We showed how to calculate the variation of population sizes in each compartment over time. A stochastic simulation implemented our model. Although there were not clinical data to validate our simulation results, a deterministic version was defined, described and also simulated with the same initial parameters as those used in stochastic simulation. The results found from each simulation were qualitatively the same (in proportion) in each compartment over time. The susceptible increased from its initial value and reached a steady state after a long period of time whereas the exposed and infectious population decreased progressively before reaching their respective steady state. Given that the birth and death rate was constant over time, the proportions of deaths and births were consequently constant over time.

However, the acceptance or refusal of statistical test results depends essentially on the distribution and significance level used. We could also have used other significance levels. For instance, if we considered $\alpha$ =0.05 then $t_{0.95}$ =1.95996. As a result, if $-t_{0.95} \leq t \leq t_{0.95}$, $H_0$ cannot be refuted. In our specific study, this change of level of significance does not have any influence on the results obtained since the number of degree of freedom is infinite. If the number of observation was not large (<120), this could have an impact on the results found.

# CHAPTER 5: CONCLUSION

Our project had as aim to work on the dynamics and transmission of malaria using Markov chains. After a brief background on simulation and modelling concepts, the dynamics of malaria and the basic model were described. This description was done at two different levels. First at individual level for hosts and vectors where we showed how a single mosquito or person is infected through mosquito bites. The results found at individual level were then extended to the population level. A deterministic model related to the stochastic model of chapter 3 was also defined. Both models, were implemented as computer simulations. With corresponding initial parameter values, the two models were run and the results obtained were compared. It was found out that the results were qualitatively the same. Quantitatively, the correlation between the data from these models was good. This was confirmed by the statistical hypothesis tests using the student t distribution at the 0.01 confidence level with infinite degrees of freedom since the number of observations were very large. The differences observed were essentially due to the random values that affected the stochastic simulation. For both models, after a long period of time we realised that the population sizes reached a steady state. The susceptible population increased from its initial value to reach a steady state. The exposed and infectious populations decreased over time and reached their respective steady state. The proportions of births and deaths were constant over time.

# References

[1] Aron, J.L. and May F.M The population Dynamics of malaria in: Roy M. Anderson (ed), the population Dynamics of Infectious Diseases: Theory and applications, Chapman and Hall, London, 1982, pp 139-179.

[2] I. Gihman, A.V Skorohod. Springer-verlag. The theory of stochastic processes I. Berlin Heidelberg New York. 1974, Volume I.

[3] I. Gihman, A.V Skorohod. Springer-verlag. The theory of stochastic processes II. Berlin Heidelberg New York. 1975, Volume II.

[4] Karlin & Taylor, A First Course in Stochastic Processes, Academic Press, 1975.

[5] Lipschutz Seymour. Probabilité. Cours et exercice. Serie Schawn. Mc Graw-Hill inc, New York, 1973.

[6] MacDonald G., 1957. The epidemiology and control of malaria Oxford. Oxford university press.

[7] Malaria, 1998. Malaria. (http://www-micro.msb.le.ac.uk/224/Malaria.html, last updated December 1998).

[8] NGWA Gideon A (2004) Modelling the dynamics of endemics malaria in growing population. Volume 4, November 2004. Page 1173-1202

[9] NGWA Gidcon A and William S.SHU, Humans, Mosquitoes and parasites in a Multi-agent Architecture for the Dynamics of Malaria. CARI 2000

[10] SPIEGEL Murray R., Probabilités et statistiques série SCHAUM.

[11] TEBOH I. Miranda: Dynamics and malaria transmission. MSc thesis Department of Mathematics and Computer Sciences. University of Buea, (April 1998)

[12] Vincent TITANJI, Julia NGU, Diane Wallace Taylor (2000) the epidemiology of plasmodium falciparum malaria in two Cameroonian villages: Simbock and Etoa. Entomological society of America.

[13] WONDJI Charles S., Didier FONTENILLE, Jean-Yves MEUNIER (2002) High malaria transmission intensity in a village close to Yaounde, the capital city of Cameroon. The American Society of Tropical and Hygiene.

# Appendix A: Algorithms

In this appendix, we present and define the algorithms related to main functions of the simulation. Note that the sentences between /* and */ or in *italic* are comments. Algorithms are coded in language C.

```
Program "sipa"          /*sipa: simulation du paludisme*/
Begin
/*Initialisation*/
read (time_step)  /*time_step ≥1: interval of time before capturing data (output results)*/
init_transition_probability( )
init_probability_distribution( )
init_population( )
init_other( )           /*birth and death rate, contact rate*/
count ← time_step

    While not end of simulation do
        While count > 0 do      /*"count" is the interval of time before
                                capturing data values. Each count might be seen
                                as a tick clock*/
            Next_transition_probability( )
            Next_probability_distribution( )

            /*Main agent activities*/
            next_compartment( )
            agent_transition( )
            Birth( )
            Bites( )
            Death( )
            count ← count - 1
        end while
    output_results( )
    count ← time_step
    end while
end
```

**- function next_transition_probability( )**

***Algorithm:*** *Multiplication of (two) matrixes:* $[p_{ij}]_{n*n} * [p_{ij}]_{n*n}$

```
Begin
    for i =1 to n do
        for j = 1 to n do
```

```
                    for k = 1 to n do
                        pij ← Σ(k=1..k=n) pik pkj
                    end for
                end for
            end for
end

- function  next_probability_distribution( )
/* multiplication vector * matrice. Π * P (see equation 4, chapter 3). Π (An array of 1
dimension) is the distribution of probability and P the transition probability matrix.
P=[pij ] n * n, Π =[π1, π2, π3, … πi, …, πn-1, πn ]
Begin
    /* At each time t*/
        for i =1 to n do
            for j = 1 to n do
                πt+1(i) ← πt(j) * pji
            end for
        end for
end
- function Agent_transition( )      /* Move agents from one compartment to
                                     another. Agents may move from compartment i
                                     to compartment k. i ∈ {S, E, I} and k∈ {S, E, I,
                                     R}
                                     */
Begin
    For each compartment (i) do
        While (population_size(i) > 0) do
            /* Choice of the next compartment of agent j */
            k ← next_compartment (agent(j))
            /* Ensure that an agent does not move to its own compartment*/
            if (k != i)    /* k different to i ?*/
                /* Remove agent j from compartment i*/
                population_size(i) ←      population_size(i) +1
                /*add agent j to compartment k */
                new_compartment(k) ← new_compartment(k)
            end if
        end while
    end for

/* Agents are effectively transferred in their new compartment*/
    for each compartment(i) do
        compartment(i) += new_compartment(i)
    end for
```

**end**

*/*Determining the compartment where an agent will be transferred. This is done randomly*/*

**- function next_compartment( )**

*/* Compartments are associated with ranges of probability values (in the range 0 to 1 inclusive). Ranges are disjoint but contiguous.*

***Principle:***

- *Obtain a probability value **and** identify the range it falls in.*
- *Identify the compartment that has this* range. */

```
Begin
n← 0          /* bad compartment*/
i←1           /*first compartment*/
      while (i ≤ number_of_compartment) and (n=0)) do
      Random(r)  /*Obtain a random (probability) value r ∈ [0  1].
                 There is not an algorithm or specification for the function Random().
                 We will use the function rand( ) provided by the programming
                 language C */
            if (r ≤ range (i))    /*Identify the range it (r) falls in*/
                  n ← i           /* Identify the compartment that has this range.*/
            end if
      i ← i+1                     /*go to next compartment*/
      end while
Return n
end
```

*/* "range" is an array that contains the $i^{th}$ row of the transition probability matrix at time t. Probability ranges are given as range(0),range(1), ... range[number_of_compartment] where the first probability range (given by range(1)) consists of all those values > than range(0) but ≤ range(1). Second range are those similarly defined by range(2). */*

**- function Birth( )**

```
Begin
      /* At each time t*/
New_Birth(t) ← X * total_population_size (t)      /* X is the birth probability to born*/
End
```

**function Death()**

```
Begin
      /* At each time t*/
Num_Death(t) ← Y(t) * total_population_size (t)      /* Y is the birth probability to die*/
End
```

**- function Bites( )**

*/* Biting is implemented as a probability value. We consider only the infectious bites. They occur if an infectious vector bites a susceptible host or if a susceptible vector bites an infectious host. Furthermore the probability to bite is considered as the probability of being exposed.*

***Variables***

*$E_h$: the probability of host being exposed*
*$E_v$: the probability of vector being exposed*
*H: total population of hosts*
*$H_s$: population of susceptible hosts*
*$H_i$: Population of infectious hosts*
*V: total population of vectors*
*$V_s$: population of susceptible vectors*
*$V_i$: Population of infectious vectors*
*$C_{hv}$: contact rate host-vector*
*$C_{vh}$: contact rate vector-host*

***Operations:***

*$E_h$ =proportion of susceptible host * proportion of infectious vector * contact rate host-vector*

*$E_v$ =proportion of susceptible vector * proportion of infectious host * contact rate vector-host */*

**Begin**

*/* At each time t*/*

$$E_h(t) \leftarrow \frac{H_s(t-1)}{H(t-1)} * \frac{V_i(t-1)}{V(t-1)} c_{hv}$$

$$E_v(t) \leftarrow \frac{V_s(t-1)}{V(t-1)} * \frac{H_i(t-1)}{H(t-1)} c_{vh}$$

**end**

# Appendix B: Initial values and output results

## B.1 Initial parameters

*Table 4. Initial values used for the simulation*

| Parameters (deterministic model) | Rates (per year) | Corresponding in the stochastic model |
|---|---|---|
| Birth rate of hosts ($\lambda_h$) | 10% | $p_{1,2}$ |
| Birth rate of vectors ($\lambda_v$) | 30% | $p_{7,8}$ |
| Death rate of hosts ($\mu_h$) | 5% | $p_{2,5}$ $p_{3,5}$ $p_{4,5}$ |
| Death rate of vectors ($\mu_v$) | 25% | $p_{8,11}$ $p_{9,11}$ $p_{10,11}$ |
| Infection rate of hosts ($v_h$) | 10% | $p_{3,4}$ |
| Infection rate of vectors ($v_v$) | 70% | $p_{9,10}$ |
| Death rate of host due to malaria ($\gamma_h$) | 1% | $p_{4,6}$ |
| Contact rate host-vector ($c_{hv}$) | 15% | $c_{hv}$ |
| Contact rate vector-host ($c_{vh}$) | 60% | $c_{vh}$ |
| Rate of exposed hosts becoming susceptible ($\phi_h$) | 30% | $p_{3,2}$ |
| Recovery rate of hosts ($r_h$) | 60% | $p_{4,2}$ |

The transition probability matrixes of hosts ($P_h$) and vectors ($P_v$) were initialised as follows:

$$P_h = \begin{array}{c} \begin{matrix} 1 & 2 & 3 & 4 & 5 & 6 \end{matrix} \\ \begin{bmatrix} 0 & 1.0 & 0 & 0 & 0 & 0 \\ 0 & 0.8 & 0.15 & 0 & 0.05 & 0 \\ 0 & 0.3 & 0.55 & 0.10 & 0.05 & 0 \\ 0 & 0.6 & 0 & 0.34 & 0.05 & 0.01 \\ 0 & 0 & 0 & 0 & 1.0 & 0 \\ 0 & 0 & 0 & 0 & 0 & 1.0 \end{bmatrix} \end{array}$$

$$P_V = \begin{array}{c} \begin{matrix} 7 & 8 & 9 & 10 & 11 \end{matrix} \\ \begin{bmatrix} 0 & 1.0 & 0 & 0 & 0 \\ 0 & 0.15 & 0.60 & 0 & 0.25 \\ 0 & 0 & 0.05 & 0.70 & 0.25 \\ 0 & 0 & 0 & 0.75 & 0.25 \\ 0 & 0 & 0 & 0 & 1.0 \end{bmatrix} \end{array}$$

For the both models, we considered that initially, 60% of hosts were susceptible, 30% exposed and 10%infectious. For the vectors 10% were susceptible, 30% exposed and 60% infectious. Hence, the initial distributions of probability were $\sigma_h$ = [0.0 0.6 0.3 0.1 0.0 0.0] for the hosts and $\sigma_v$ = [0.0 0.1 0.3 0.6 0.0] for the vectors. Initially, $N_h(0)$ = 1,000,000 and $N_v(0)$ = 10,000,000. This means that for the hosts, $S_h(0)$ = 600,000, $E_h(0)$ = 300,000 and $I_h(0)$= 100,000. For the vectors, $S_v(0)$ = 1,000,000, $E_v(0)$=3,000,000 and $I_v(0)$ = 6,000,000.

Note that, the unit time used was a day. Since the different rates were expressed per annum (except the contact rate), for the simulation purposes, each rate was divided into 365 equal parts corresponding to each day. Thus, during the simulation process, the number of subjects depending on a given rate was calculated per day in each compartment. In the stochastic simulation, those rates were considered as transition probabilities.

### B.2 Outputs

For the deterministic simulation, the file "theopop.txt" contains outputs that include population sizes in each compartment. For the stochastic simulation there are six output filers; "poph.txt" and "popv.txt" contain the number of subjects in each state for the hosts and vectors respectively. "outputh.txt" and "outputv.txt" contain the transition probabilities in the host and vector compartments respectively. "disprobh.txt" and "disprobv.txt" contain the distribution of probabilities in the host and vector compartments respectively. Simulation is automatically stopped if a population (host or vector) disappears.

### B.2) Fractional population of hosts

| | Stochastic simulation | | | | | |
|---|---|---|---|---|---|---|
| Time step | Bh(M) | Sh(M) | Eh(M) | Ih(M) | Ro(M) | Ri(M) |
| 0 | 0 | 0.6 | 0.3 | 0.1 | 0 | 0 |
| 1 | 0.086128943 | 0.59945374 | 0.29877203 | 0.09956371 | 0.00427295 | 2.70961E-05 |
| 2 | 0.086129077 | 0.59989207 | 0.29854826 | 0.09934993 | 0.00427351 | 2.58589E-05 |
| 3 | 0.086129065 | 0.60047718 | 0.2981764 | 0.09913648 | 0.00427366 | 2.58171E-05 |
| 4 | 0.086129037 | 0.60088953 | 0.29797726 | 0.09892366 | 0.00427342 | 2.55771E-05 |
| 5 | 0.086128988 | 0.6016647 | 0.29741451 | 0.09871105 | 0.00427357 | 2.57338E-05 |
| 6 | 0.086129056 | 0.60243859 | 0.296853 | 0.09849928 | 0.00427313 | 2.54946E-05 |
| 7 | 0.086128969 | 0.60321056 | 0.29629223 | 0.0982877 | 0.00427367 | 2.54534E-05 |
| 8 | 0.086128997 | 0.60398138 | 0.29573276 | 0.09807656 | 0.00427323 | 2.56093E-05 |
| 9 | 0.086129137 | 0.60475048 | 0.29517423 | 0.09786582 | 0.00427358 | 2.53713E-05 |
| 10 | 0.086128993 | 0.60551831 | 0.29461655 | 0.09765571 | 0.00427353 | 2.53303E-05 |
| ... | ... | ... | ... | ... | ... | ... |
| 1482 | 0.086129033 | 0.96158009 | 0.02687336 | 0.00936023 | 0.00427346 | 2.38182E-06 |
| 1483 | 0.086129033 | 0.96168739 | 0.02678354 | 0.0093428 | 0.00427343 | 2.36017E-06 |
| 1484 | 0.086129033 | 0.96174035 | 0.0267559 | 0.00931745 | 0.00427346 | 2.35631E-06 |
| 1485 | 0.086129033 | 0.96183309 | 0.02668214 | 0.00929849 | 0.00427343 | 2.35245E-06 |
| 1486 | 0.086129033 | 0.9619142 | 0.02662423 | 0.00927528 | 0.00427346 | 2.34859E-06 |
| 1487 | 0.086129033 | 0.96198909 | 0.02657334 | 0.00925132 | 0.00427345 | 2.32711E-06 |
| 1488 | 0.086129032 | 0.96206089 | 0.02652253 | 0.0092303 | 0.00427346 | 2.3409E-06 |
| 1489 | 0.086129033 | 0.96207767 | 0.02652575 | 0.00921033 | 0.00427344 | 2.3195E-06 |
| 1490 | 0.086129034 | 0.96215785 | 0.02645877 | 0.00919714 | 0.00427345 | 2.33324E-06 |
| 1491 | 0.086129033 | 0.96201915 | 0.02661568 | 0.00917892 | 0.00427345 | 2.3119E-06 |
| 1492 | 0.086129033 | 0.96215598 | 0.02645665 | 0.00920111 | 0.00427346 | 2.3256E-06 |
| 1493 | 0.086129032 | 0.96223989 | 0.02638215 | 0.0091917 | 0.00427345 | 2.32179E-06 |
| 1494 | 0.086129033 | 0.96234003 | 0.02630069 | 0.00917302 | 0.00427344 | 2.33541E-06 |
| 1495 | 0.086129032 | 0.96242695 | 0.02623804 | 0.00914875 | 0.00427346 | 2.31419E-06 |
| 1496 | 0.086129033 | 0.96232675 | 0.0263636 | 0.00912345 | 0.00427344 | 2.29303E-06 |
| 1497 | 0.086129033 | 0.96245477 | 0.02622469 | 0.00913429 | 0.00427346 | 2.30661E-06 |
| 1498 | 0.086129033 | 0.96253247 | 0.02616292 | 0.00911835 | 0.00427345 | 2.32015E-06 |
| 1499 | 0.086129033 | 0.96263123 | 0.02608386 | 0.00909868 | 0.00427345 | 2.29906E-06 |
| 1500 | 0.086129033 | 0.96258975 | 0.02615061 | 0.00907344 | 0.00427343 | 2.29529E-06 |

| | Deterministic simulation | | | | | |
|---|---|---|---|---|---|---|
| Time step | Bh(D) | Sh(D) | Eh(D) | Ih(D) | Ro(D) | Ri(D) |
| 0 | 0 | 0.6 | 0.3 | 0.1 | 0 | 0 |
| 1 | 0.086937651 | 0.55425454 | 0.26403988 | 0.08668455 | 0.004522747 | 4.33403E-05 |
| 2 | 0.086937641 | 0.5652045 | 0.25673349 | 0.08304259 | 0.004522756 | 4.15194E-05 |
| 3 | 0.086937652 | 0.57585617 | 0.24956245 | 0.07956355 | 0.004522763 | 3.97799E-05 |
| 4 | 0.08693763 | 0.58622635 | 0.24251773 | 0.07623962 | 0.004522771 | 3.8118E-05 |
| 5 | 0.086937658 | 0.59632904 | 0.23559267 | 0.07306343 | 0.004522778 | 3.65299E-05 |
| 6 | 0.086937614 | 0.60617627 | 0.2287824 | 0.07002785 | 0.004522785 | 3.50121E-05 |
| 7 | 0.086937598 | 0.61577798 | 0.2220837 | 0.06712616 | 0.004522792 | 3.35613E-05 |
| 8 | 0.08693762 | 0.62514276 | 0.21549437 | 0.06435198 | 0.004522797 | 3.21743E-05 |
| 9 | 0.086937653 | 0.63427754 | 0.20901359 | 0.06169916 | 0.004522802 | 3.08479E-05 |
| 10 | 0.086937598 | 0.64318848 | 0.20264096 | 0.05916196 | 0.00452281 | 2.95793E-05 |
| ... | ... | ... | ... | ... | ... | ... |
| 1482 | 0.086937591 | 0.9050182 | 1.0989E-37 | 4.2959E-34 | 0.004522943 | 2.14803E-37 |
| 1483 | 0.086937639 | 0.9050182 | 1.0372E-37 | 4.0804E-34 | 0.004522944 | 2.04026E-37 |
| 1484 | 0.086937636 | 0.90501821 | 9.7896E-38 | 3.8757E-34 | 0.004522941 | 1.93791E-37 |
| 1485 | 0.086937644 | 0.9050182 | 9.2401E-38 | 3.6812E-34 | 0.004522943 | 1.84068E-37 |
| 1486 | 0.086937607 | 0.90501821 | 8.7215E-38 | 3.4965E-34 | 0.004522942 | 1.74834E-37 |
| 1487 | 0.08693764 | 0.9050182 | 8.232E-38 | 3.3211E-34 | 0.004522942 | 1.66063E-37 |
| 1488 | 0.086937573 | 0.90501821 | 7.7699E-38 | 3.1545E-34 | 0.004522942 | 1.57731E-37 |
| 1489 | 0.086937627 | 0.9050182 | 7.3338E-38 | 2.9962E-34 | 0.004522942 | 1.49818E-37 |
| 1490 | 0.08693764 | 0.9050182 | 6.9221E-38 | 2.8459E-34 | 0.004522943 | 1.42302E-37 |
| 1491 | 0.086937614 | 0.90501821 | 6.5336E-38 | 2.7032E-34 | 0.004522942 | 1.35163E-37 |
| 1492 | 0.086937599 | 0.9050182 | 6.1669E-38 | 2.5675E-34 | 0.004522942 | 1.28382E-37 |
| 1493 | 0.086937647 | 0.90501821 | 5.8207E-38 | 2.4387E-34 | 0.004522943 | 1.21941E-37 |
| 1494 | 0.086937608 | 0.9050182 | 5.494E-38 | 2.3164E-34 | 0.004522943 | 1.15824E-37 |
| 1495 | 0.08693763 | 0.9050182 | 5.1856E-38 | 2.2002E-34 | 0.004522943 | 1.10013E-37 |
| 1496 | 0.086937618 | 0.90501821 | 4.8945E-38 | 2.0898E-34 | 0.004522942 | 1.04494E-37 |
| 1497 | 0.08693762 | 0.9050182 | 4.6198E-38 | 1.9849E-34 | 0.004522943 | 9.92513E-38 |
| 1498 | 0.086937589 | 0.90501821 | 4.3605E-38 | 1.8854E-34 | 0.004522942 | 9.4272E-38 |
| 1499 | 0.086937614 | 0.9050182 | 4.1157E-38 | 1.7908E-34 | 0.004522942 | 8.95425E-38 |
| 1500 | 0.086937608 | 0.90501821 | 3.8847E-38 | 1.7009E-34 | 0.004522942 | 8.50502E-38 |

## B.3) Fractional population of vectors

| | Stochastic simulation | | | | | |
|---|---|---|---|---|---|---|
| Time step | Bh(M) | Sh(M) | Eh(M) | Ih(M) | Ro(M) | Ri(M) |
| 0 | 0 | 0.6 | 0.3 | 0.1 | 0 | 0 |
| 1 | 0.086128943 | 0.59945374 | 0.29877203 | 0.09956371 | 0.00427295 | 2.70961E-05 |
| 2 | 0.086129077 | 0.59989207 | 0.29854826 | 0.09934993 | 0.00427351 | 2.58589E-05 |
| 3 | 0.086129065 | 0.60047718 | 0.2981764 | 0.09913648 | 0.00427366 | 2.58171E-05 |
| 4 | 0.086129037 | 0.60088953 | 0.29797726 | 0.09892366 | 0.00427342 | 2.55771E-05 |
| 5 | 0.086128988 | 0.6016647 | 0.29741451 | 0.09871105 | 0.00427357 | 2.57338E-05 |
| 6 | 0.086129056 | 0.60243859 | 0.296853 | 0.09849928 | 0.00427313 | 2.54946E-05 |
| 7 | 0.086128969 | 0.60321056 | 0.29629223 | 0.0982877 | 0.00427367 | 2.54534E-05 |
| 8 | 0.086128997 | 0.60398138 | 0.29573276 | 0.09807656 | 0.00427323 | 2.56093E-05 |
| 9 | 0.086129137 | 0.60475048 | 0.29517423 | 0.09786582 | 0.00427358 | 2.53713E-05 |
| 10 | 0.086128993 | 0.60551831 | 0.29461655 | 0.09765571 | 0.00427353 | 2.53303E-05 |
| ... | ... | ... | ... | ... | ... | ... |
| 1482 | 0.086129033 | 0.96158009 | 0.02687336 | 0.00936023 | 0.00427346 | 2.38182E-06 |
| 1483 | 0.086129033 | 0.96168739 | 0.02678354 | 0.0093428 | 0.00427343 | 2.36017E-06 |
| 1484 | 0.086129033 | 0.96174035 | 0.0267559 | 0.00931745 | 0.00427346 | 2.35631E-06 |
| 1485 | 0.086129033 | 0.96183309 | 0.02668214 | 0.00929849 | 0.00427343 | 2.35245E-06 |
| 1486 | 0.086129033 | 0.9619142 | 0.02662423 | 0.00927528 | 0.00427346 | 2.34859E-06 |
| 1487 | 0.086129033 | 0.96198909 | 0.02657334 | 0.00925132 | 0.00427345 | 2.32711E-06 |
| 1488 | 0.086129032 | 0.96206089 | 0.02652253 | 0.0092303 | 0.00427346 | 2.3409E-06 |
| 1489 | 0.086129033 | 0.96207767 | 0.02652575 | 0.00921033 | 0.00427344 | 2.3195E-06 |
| 1490 | 0.086129034 | 0.96215785 | 0.02645877 | 0.00919714 | 0.00427345 | 2.33324E-06 |
| 1491 | 0.086129033 | 0.96201915 | 0.02661568 | 0.00917892 | 0.00427345 | 2.3119E-06 |
| 1492 | 0.086129033 | 0.96215598 | 0.02645665 | 0.00920111 | 0.00427346 | 2.3256E-06 |
| 1493 | 0.086129032 | 0.96223989 | 0.02638215 | 0.0091917 | 0.00427345 | 2.32179E-06 |
| 1494 | 0.086129033 | 0.96234003 | 0.02630069 | 0.00917302 | 0.00427344 | 2.33541E-06 |
| 1495 | 0.086129032 | 0.96242695 | 0.02623804 | 0.00914875 | 0.00427346 | 2.31419E-06 |
| 1496 | 0.086129033 | 0.96232675 | 0.0263636 | 0.00912345 | 0.00427344 | 2.29303E-06 |
| 1497 | 0.086129033 | 0.96245477 | 0.02622469 | 0.00913429 | 0.00427346 | 2.30661E-06 |
| 1498 | 0.086129033 | 0.96253247 | 0.02616292 | 0.00911835 | 0.00427345 | 2.32015E-06 |
| 1499 | 0.086129033 | 0.96263123 | 0.02608386 | 0.00909868 | 0.00427345 | 2.29906E-06 |
| 1500 | 0.086129033 | 0.96258975 | 0.02615061 | 0.00907344 | 0.00427343 | 2.29529E-06 |

| | Deterministic simulation | | | | | |
|---|---|---|---|---|---|---|
| Time step | Bh(D) | Sh(D) | Eh(D) | Ih(D) | Ro(D) | Ri(D) |
| 0 | 0 | 0.6 | 0.3 | 0.1 | 0 | 0 |
| 1 | 0.086937651 | 0.55425454 | 0.26403988 | 0.08668455 | 0.004522747 | 4.33403E-05 |
| 2 | 0.086937641 | 0.5652045 | 0.25673349 | 0.08304259 | 0.004522756 | 4.15194E-05 |
| 3 | 0.086937652 | 0.57585617 | 0.24956245 | 0.07956355 | 0.004522763 | 3.97799E-05 |
| 4 | 0.08693763 | 0.58622635 | 0.24251773 | 0.07623962 | 0.004522771 | 3.8118E-05 |
| 5 | 0.086937658 | 0.59632904 | 0.23559267 | 0.07306343 | 0.004522778 | 3.65299E-05 |
| 6 | 0.086937614 | 0.60617627 | 0.2287824 | 0.07002785 | 0.004522785 | 3.50121E-05 |
| 7 | 0.086937598 | 0.61577798 | 0.2220837 | 0.06712616 | 0.004522792 | 3.35613E-05 |
| 8 | 0.08693762 | 0.62514276 | 0.21549437 | 0.06435198 | 0.004522797 | 3.21743E-05 |
| 9 | 0.086937653 | 0.63427754 | 0.20901359 | 0.06169916 | 0.004522802 | 3.08479E-05 |
| 10 | 0.086937598 | 0.64318848 | 0.20264096 | 0.05916196 | 0.00452281 | 2.95793E-05 |
| ... | ... | ... | ... | ... | ... | ... |
| 1482 | 0.086937591 | 0.9050182 | 1.0989E-37 | 4.2959E-34 | 0.004522943 | 2.14803E-37 |
| 1483 | 0.086937639 | 0.9050182 | 1.0372E-37 | 4.0804E-34 | 0.004522944 | 2.04026E-37 |
| 1484 | 0.086937636 | 0.90501821 | 9.7896E-38 | 3.8757E-34 | 0.004522941 | 1.93791E-37 |
| 1485 | 0.086937644 | 0.9050182 | 9.2401E-38 | 3.6812E-34 | 0.004522943 | 1.84068E-37 |
| 1486 | 0.086937607 | 0.90501821 | 8.7215E-38 | 3.4965E-34 | 0.004522942 | 1.74834E-37 |
| 1487 | 0.08693764 | 0.9050182 | 8.232E-38 | 3.3211E-34 | 0.004522942 | 1.66063E-37 |
| 1488 | 0.086937573 | 0.90501821 | 7.7699E-38 | 3.1545E-34 | 0.004522942 | 1.57731E-37 |
| 1489 | 0.086937627 | 0.9050182 | 7.3338E-38 | 2.9962E-34 | 0.004522942 | 1.49818E-37 |
| 1490 | 0.08693764 | 0.9050182 | 6.9221E-38 | 2.8459E-34 | 0.004522943 | 1.42302E-37 |
| 1491 | 0.086937614 | 0.90501821 | 6.5336E-38 | 2.7032E-34 | 0.004522942 | 1.35163E-37 |
| 1492 | 0.086937599 | 0.9050182 | 6.1669E-38 | 2.5675E-34 | 0.004522942 | 1.28382E-37 |
| 1493 | 0.086937647 | 0.90501821 | 5.8207E-38 | 2.4387E-34 | 0.004522943 | 1.21941E-37 |
| 1494 | 0.086937608 | 0.9050182 | 5.494E-38 | 2.3164E-34 | 0.004522943 | 1.15824E-37 |
| 1495 | 0.08693763 | 0.9050182 | 5.1856E-38 | 2.2002E-34 | 0.004522943 | 1.10013E-37 |
| 1496 | 0.086937618 | 0.90501821 | 4.8945E-38 | 2.0898E-34 | 0.004522942 | 1.04494E-37 |
| 1497 | 0.08693762 | 0.9050182 | 4.6198E-38 | 1.9849E-34 | 0.004522943 | 9.92513E-38 |
| 1498 | 0.086937589 | 0.90501821 | 4.3605E-38 | 1.8854E-34 | 0.004522942 | 9.4272E-38 |
| 1499 | 0.086937614 | 0.9050182 | 4.1157E-38 | 1.7908E-34 | 0.004522942 | 8.95425E-38 |
| 1500 | 0.086937608 | 0.90501821 | 3.8847E-38 | 1.7009E-34 | 0.004522942 | 8.50502E-38 |

# Appendix C: Sample implementation code

## C.1 Stochastic model

**int nextTransitionProbabilities(int nComp,int nCompV)**

```
{
int i,j,k;
float sum;
 for (i=1;i<=nComp;i++)
 {  for (j=1;j<=nComp;j++)
    {  sum=0.0;
       for (k=1;k<=nComp;k++)
        {
          sum+=pth[i][k]*oldpth[k][j];
        }
      oldpth[i][j] = newpth[i][j] = sum;
      }
}
 for (i=1;i<=nCompV;i++)
 {  for (j=1;j<=nCompV;j++)
    {  sum=0.0;
       for (k=1;k<=nCompV;k++)
        {
         sum+=ptv[i][k]*oldptv[k][j];
        }
     oldptv[i][j] = newptv[i][j] = sum;
     }
  }
return 0;
}
/*++++++++++++++++++++++++++++++++++++++++++++++++++++++++++
+++++*/
```

**int nextProbabilitiesDistribution(int nComp,int nCompV)**

```
{
int i,j,m;
float sum;
        for (i=1;i<=nComp;i++)
        { sum=0.0;
                for (j=1;j<=nComp;j++)
                { sum+= p[j]*newpth[j][i];
                }
        newp[i]=sum;
```

```
        }
        for (i=1;i<=nCompV;i++)
        {  sum=0.0;
                for (j=1;j<=nCompV;j++)
                { sum+= q[j]*newptv[j][i];
                }
                 newq[i]=sum;
        }
return 0;
}
/**+++++++++++++++++++++++++++++++++++++++++++++++++++++++++
+++++++++++*/
int birth(int nComp,int nCompV )
{
int i;
long new_BornH = 0;
long new_BornV = 0;
Proba_of_Birth_H = Birth_Rate_H / Year;
num_BH += (popC[0] * Proba_of_Birth_H);  /*Calcul des nombre de naissance chez les hôtes */
if (num_BH >= 1.0)
{
  while (num_BH > 1.0 )
  {
  new_BornH++;
  num_BH--;
  }
  popC[B_COMP] += new_BornH;   /*les nouveaux sont envoyés dans leur compartiment (B_COMP)*/
  popC[S_COMP] += new_BornH;  /* Les nouveaux nés sont supposés susceptible*/
  new_BornH = 0;
  }
/*NAISSANCES CHEZ LES VECTEURS*/
Proba_of_Birth_V = Birth_Rate_V / Month;
num_BV += popCV[0] * Proba_of_Birth_V;
if (num_BV >= 1.0)
{
  while (num_BV > 1.0 )
  {
  new_BornV++;
  num_BV--;
  }
  popCV[B_COMPV] += new_BornV; /*les nouveaux sont envoyés dans leur compartiment (B_COMPV)*/
```

```
  popCV[S_COMPV] += new_BornV; /* Les nouveaux n,s sont supposés
susceptible*/
  new_BornV = 0;
  }
popC[0] = popCV[0] =0;
for(i=S_COMP; i<= Host_Transient_State; i++) popC[0] += popC[i];
for(i=S_COMPV; i<= Vector_Transient_State; i++) popCV[0] +=popCV[i];
for(i=0; i<= nComp; i++) old_popC[i] =popC[i];
for(i=0; i<= nCompV; i++) old_popCV[i] =popCV[i];
return 0;
}

/*++++++++++++++++++++++++++++++++++++++++++++++++++++++++++++
+++++*/
int death(int nComp, int nCompV)
{
int i;
long Num_Death = 0; /*Comptabilise les morts par unit, de temps*/
num_Iv= popCV[I_COMPV];
if (popC[I_COMP]>0) tDeath_HI += (popC[I_COMP] * (newp[I_COMP]/Year));

Proba_to_die_HO = Death_Rate_HO / Year;
 for (i=S_COMP; i <= Host_Transient_State; i++)
 {
 if (popC[i]>0)num_death_HO[i] += (popC[i] * Proba_to_die_HO); /* les morts sont
recens,s dans chaque compartiment*/
 }
for (i=S_COMP; i <= Host_Transient_State; i++)
 {
   if (num_death_HO[i] >= 1.0)
   {
   while (num_death_HO[i] > 1.0)
   {
   Num_Death++;
   num_death_HO[i]--;
   }
   popC[i] -= Num_Death;
   popC[RO_COMP] += Num_Death;
   Num_Death = 0;
   }
 }
if (tDeath_HI >= 1.0)
   {
   while (tDeath_HI > 1.0)
   {
```

```
  Num_Death++;
  tDeath_HI--;
  }
  popC[I_COMP] -= Num_Death;            /* des personnes meurent de suite de
malaria.*/
  popC[RI_COMP] += Num_Death; /*les personnes mortes de suite de malaria sont
de*/
  Num_Death = 0;
  }
Proba_to_die_V = Death_Rate_V  / Month;
 for (i=S_COMPV; i <= Vector_Transient_State; i++)
 {
 if (popCV[i]>0)
 num_death_V[i] += (popCV[i] * Proba_to_die_V);
 }
for (i=S_COMPV; i <= Vector_Transient_State; i++)
 {
  if (num_death_V[i] >= 1.0)
  {
  while (num_death_V[i] > 1.0)
  {
  Num_Death++;
  num_death_V[i]--;
  if (i==I_COMPV) {mortV++; }
  }
  popCV[i] -= Num_Death;
  popCV[R_COMPV] += Num_Death;
  Num_Death = 0;
  }
 }
popC[0] = popCV[0] = 0;
for(i = S_COMP; i<= Host_Transient_State; i++) popC[0] +=popC[i];
for(i = S_COMPV; i<= Vector_Transient_State; i++) popCV[0] +=popCV[i];
for(i = 0; i<= nComp; i++) old_popC[i] =popC[i];
for(i = 0; i<= nCompV; i++) old_popCV[i] =popCV[i];
return 0;
}
/*++++++++++++++++++++++++++++++++++++++++++++++++++++++++++
+++++*/
int nextComp(int nComp)
{
 int i, n;
 float r, r1, r2;
 n = BAD_COMPARTMENT;
 r1 = (rand() * 1.0) / RAND_MAX;
```

```
 r2 = (rand() % Host_Transient_State-1)+2;
 r =r1 + r2;
 i = S_COMP;
 do
   if (r <= rangeH[i]) {
     n = i;
     break;
   }
 while (++i <= nComp-Host_Absorbing_State);
 return n;
}
/*++++++++++++++++++++++++++++++++++++++++++++++++++++++++
+++++*/
int nextCompV(int nCompV)
{
 int i, n;
 float r, r1, r2;
 n = BAD_COMPARTMENT;
 r1 = (rand() * 1.0) / RAND_MAX;
 r2 = (rand() % (Vector_Transient_State - stateV + 1)) + stateV;
        r =r1 + r2;
 i = stateV;
 do
   if (r <= rangeV[i]) {
     n = i;
     break;
   }
 while (++i <= nCompV-Vector_Absorbing_State );    /*l'etat  absorbant est exclus*/
 return n;
}
/*++++++++++++++++++++++++++++++++++++++++++++++++++++++++
+++++*/
int bite(int nComp, int nCompV)
{
int i,j,k;
if (popC[0]==0)
{
 MessageDlg("THERE IS NOT MORE POPULATION IN HOST
COMPARTMENT\nEND OF EXECUTION DUE TO LACK OF POPULATION",
mtInformation, TMsgDlgButtons() << mbOK, 0);
 Pop_Exist= 0;
}
else
if (popCV[0]<=0)
```

```
{ MessageDlg("THERE IS NOT MORE POPULATION IN VECTOR COMPARTMENT\n END OF EXECUTION DUE TO LACK OF POPULATION", mtInformation, TMsgDlgButtons() << mbOK, 0);

Pop_Exist=0;
}
else

oldpth[S_COMPV][E_COMPV] = newpth[S_COMP][E_COMP] = (popC[S_COMP] * popCV[I_COMPV]) / (popC[nComp-nComp] * popCV[nCompV-nCompV]) * chv;
oldptv[S_COMPV][E_COMPV] = newptv[S_COMPV][E_COMPV] = (popC[I_COMP] * popCV[S_COMPV]) / (popC[0]*popCV[0]) * cvh;
}
return 0;
}
/*++++++++++++++++++++++++++++++++++++++++++++++++++++++++++++++++
+++++*/
int agentTransition(int nComp, int nCompV)
/* move agents from one compartment to another */
{
int i,j,k;
long x,y;
long int tpopH[MAX_COMP];
long int tpopV[MAX_COMP];
long int tpopH1[MAX_COMP];
long int tpopV1[MAX_COMP];
for (i=0; i<= nComp-2; i++) tpopH1[i]=0;
for (i=0; i<= nCompV-1; i++)tpopV1[i]=0;

for (i= RO_COMP; i<= nComp; i++)  tpopH[i]=popC[i] ;
for (i = S_COMPV; i <= nCompV; i++ ) old_popCV[i] = popCV[i];
for (i = S_COMP; i <= Host_Transient_State; i++ ) old_popC[i] = popC[i];
tpopV[R_COMPV]=popCV[R_COMPV] ;
 for (j = S_COMP; j <= Host_Transient_State; j++ )          /* execute for each compartment */
  {
   state=j;                                  /*mémorisation de l'état courant.*/
   initRanges(nComp);                        /* initialise probability ranges */
    for (i= S_COMP; i<= Host_Transient_State; i++) tpopH[i] = 0;     /* initialise popn increments */
        x=popC[j];
    for (k=x; k>0;k--) {
     i = nextComp(nComp);                    /* new compartment for agent k */
```

```
    if ((j==S_COMP)&&(i==I_COMP)) i = BAD_COMPARTMENT;
    if ((j==I_COMP)&&(i==E_COMP)) i = BAD_COMPARTMENT;
    if (i == BAD_COMPARTMENT)
    {
       i = j;                              /* ignore invalid transition */
    }
    if (i != j) {                                  /* ignore transition to same comp. */
       --popC[j];                          /* decrement k's current compartment */
       ++tpopH[i];
       ++tpopH1[i];
                 }
  }
  if(old_popC[state] > 0){
    for (k = S_COMP; k <= nComp ; k++){
    if (k != state)
    {
     oldpth[state][k] = newpth[state][k] = (tpopH[k] * 1.0) / (old_popC[state]
+tpopH[RO_COMP]+tpopH[RI_COMP]);
    }
    else
    {
     oldpth[state][k]=newpth[state][k]=(popC[state]          *          1.0)          /
(old_popC[state]+tpopH[RO_COMP]+tpopH[RI_COMP]);
    }
    }
 }
    }
  for (j=S_COMPV; j <= Vector_Transient_State-1 ; j++ )            /*   execute   for
each compartment */
 {
  stateV=j;                                                  /*mémorisation        de
l'état courant.*/
  initRangesV(nCompV);                                       /* initialise probability
ranges */
   for (i= 0; i<=  Vector_Transient_State ; i++) tpopV[i] = 0;            /* initialise popn
increments */
 /*  for (i = S_COMPV; i <= nCompV; i++ ) old_popCV[i] = popCV[i];
     y=popCV[stateV];
   for (k= y ; k>0; k--) {
    i = nextCompV(nCompV);                   /* new compartment for agent k */
    if ((j==S_COMPV)&&(i==I_COMPV)) i= BAD_COMPARTMENT;
    if (i == BAD_COMPARTMENT)
    {
       i = j;                              /* ignore invalid transition */
    }
```

```
    if (i != j) {                                   /* ignore transition to same comp. */
       --popCV[j];                          /* decrement k's current compartment */
       ++tpopV[i];                          /* increment k's new compartment */
       ++tpopV1[i];
    }
   }
 if (old_popCV[stateV] > 0){
  for (k = stateV; k <=nCompV; k++){
   if (k != stateV)
   {
   oldptv[stateV][k] = newptv[stateV][k] = (tpopV[k] * 1.0) / (old_popCV[stateV] +
tpopV[R_COMPV]);
   }
   else
   {
   oldptv[stateV][k]   =   newptv[stateV][k]   =     (popCV[stateV]   *   1.0)   /
(old_popCV[stateV] + tpopV[R_COMPV]);
   }
 }
   }
  }
  for (i= S_COMPV; i<= nCompV-1; i++)   popCV[i] += tpopV1[i];  /*               add
increments to compartments */
  for (i= S_COMP; i<= Host_Transient_State; i++) popC[i] += tpopH1[i];
        /* add increments to compartments */
  return 0;
}
//---------------------------------------------------------------------------
void __fastcall TForm4::Timer1Timer(TObject *Sender)
{
   srand((unsigned int)time('\0'));
if (arret==1)
{
fclose(outputPop);
  fclose(outputpH);
  fclose(outputPopV);
  fclose(outputpV);
  fclose(outputDisprobH);
  fclose(outputDisprobV);
  Button1->Enabled=true;
  Form3->Button3->Enabled=false;
  Form2->Show();
  Timer3->Enabled =false;
  Timer1->Enabled=false;
  return;
```

```
}
     Timer1->Tag++;;
   if (Timer1->Tag > val)
   {
   Form3->Button3->Enabled=false;
   Form2->Show();
   Timer3->Enabled =false;
   Timer1->Enabled=false;
   Button1->Enabled=true;
   return;
   }
 if (Timer1->Tag <= val)
{
 timeStep_cur=0;
  fprintf(outputPop,"\n\t === Year %d === \n", Timer1->Tag);
  fprintf(outputPopV,"\n\t === Year %d === \n", Timer1->Tag);
  fprintf(outputDisprobV,"\n\t === Year %d === \n", Timer1->Tag);
  fprintf(outputDisprobH,"\n\t === Year %d === \n", Timer1->Tag);
  fprintf(outputPop,"\n");
  fprintf(outputPopV,"\n");
  fprintf(outputDisprobV,"\n");
  fprintf(outputDisprobH,"\n");
    counter = timeStep_skip;                    /*   initialise/reset   down   counter   for
timeSteps to skip */
    if (!IsValid)
    {
    Timer1->Enabled=false;
    Timer2->Enabled=false;
    return;
    }
Timer1->Enabled=false;
Timer2->Enabled=true;
}}
//------------------------------------------------------------------------
void __fastcall TForm4::Timer2Timer(TObject *Sender)
{
if (arret==1)
{
fclose(outputPop);
 fclose(outputpH);
 fclose(outputPopV);
 fclose(outputpV);
 fclose(outputDisprobH);
 fclose(outputDisprobV);
 Button1->Enabled=true;
```

```
 Form3->Button3->Enabled=false;
 Form2->Show();
 Timer3->Enabled =false;
 Timer2->Enabled=false;
 return;
}
if(!Pop_Exist)
            {
         // Form3->Button2->Enabled =false;
          Form3->Button3->Enabled =false;
          Form3->Button4->Enabled =false;
          Timer1->Enabled=false;
          Timer2->Enabled=false;
            return;
            }
if (Form3->alertepause==1 )
{ Form4->Button1->Enabled=true;
Form3->alertepause=0;
if (Timer1->Enabled ==true ) {ancienvrait1=1;Timer1->Enabled =false; }
if (Timer2->Enabled ==true ) {ancienvrait2=1;Timer2->Enabled =false; }
//Form3->Button2->Enabled=true;
Form3->Button4->Enabled=true;
Form3->Button3->Enabled=false;
}
 if(timeStep_cur++ > timeStep_max)
 {
  Timer2->Enabled=false;
  Timer1->Enabled=true;
  timeStep_cur =0;
 }
 else
 {
       if(!Pop_Exist)
            {
         // Form3->Button2->Enabled =false;
          Form3->Button3->Enabled =false;
          Form3->Button4->Enabled =false;
          Timer1->Enabled=false;
          Timer2->Enabled=false;
            return;
            }
  else
  {
      birth(nComp,nCompV);
      bite(nComp, nCompV);
```

```
        death(nComp, nCompV);
      agentTransition(nComp,nCompV);
      nextTransitionProbabilities(nComp,nCompV);
      nextProbabilitiesDistribution(nComp,nCompV);
   }
     if (!(--counter)) {                      /* print only when cnt is zero */
        counter = timeStep_skip;              /* reset down counter */
        outData(nComp,nCompV);                /* output popn & transition probs in host
compartment */
        popCV[R_COMPV]=popC[RI_COMP]=popC[RO_COMP]=popCV[B_COM
PV]=popC[B_COMP]=0;
     }
}}
//---------------------------------------------------------------------------
```

**void __fastcall TForm4::Button2Click(TObject *Sender)**

```
{
arret=1;
//Form3->Button2->Enabled =false;
Form3->Button3->Enabled =false;
Form3->Button4->Enabled =false;
Button1->Enabled=true;
}
//---------------------------------------------------------------------------
```

**void __fastcall TForm4::Timer3Timer(TObject *Sender)**

```
{
//*********************CONTINUE
if ( Form3->alertecontinue ==1)
{  //Form3->Button2->Enabled=false;
 Form3->alertecontinue=0;
if (ancienvrait1==1) {Timer1->Enabled =true; ancienvrait1=0;}
if (ancienvrait2==1) {Timer2->Enabled =true; ancienvrait2=0;}
Form3->Button3->Enabled=true;
//Form3->Button2->Enabled=false;
Form3->Button4->Enabled=false;
}}
```

**C.2 Deterministic model**

**int func_prime (int size)**

```
{
Nh = oldy[1] + oldy[2] + oldy[3];                /*population of host*/
Nv = oldy[4] + oldy[5] + oldy[6];                /*population of host*/
yprime[1]                          =lambdaH*Nh+r_h*oldy[3]+oldy[2]*psi-
((alphaH+cvh1*oldy[6]/Nh)*oldy[1]); /*sucepible host*/
```

```
yprime[2] = cvh1*oldy[6]*oldy[1]/Nh -((alphaH + psi+betaH)*oldy[2]);     /*exposed
host*/
yprime[3] = betaH*oldy[2] - ((alphaH + r_h + gamma)*oldy[3]);
        /*infectious host*/
yprime[4] = Nv*lambdaV- ((alphaV + cvh1*oldy[3]/Nh) * oldy[4]);
        /*susceptible vector*/
yprime[5] = chv1 * oldy[3]*oldy[4]/Nh- ((alphaV + betaV) *oldy[5]);
        /*exposed vector*/
yprime[size1] = betaV*oldy[5]- alphaV *oldy[6];                    /*infectious
vector*/
ynew[7] = lambdaH * Nh;
ynew[8] = alphaH*Nh;
ynew[9] =  gamma*ynew[3];
ynew[10] =  Nv*lambdaV;
ynew[11] = alphaV*Nv;
return 0;
}

/*++++++++++++++++++++++++++++++++++++++++++++++++++++++++++++++++++++
+++++*/
int equa_diff (int size)
{
int i;
func_prime(size);
for (i=1; i<= size; i++) ynew[i] = oldy[i] +h*yprime[i];
for (i=1; i<=size; i++) oldy[i] = ynew[i];
for (i=1; i<=size; i++) oldy[i] = ynew[i];
return 0;
}
```

www.ingramcontent.com/pod-product-compliance
Lightning Source LLC
LaVergne TN
LVHW022312200926
848481LV00004B/1298